THE HISTORY OF
SPACE
VEHICLES

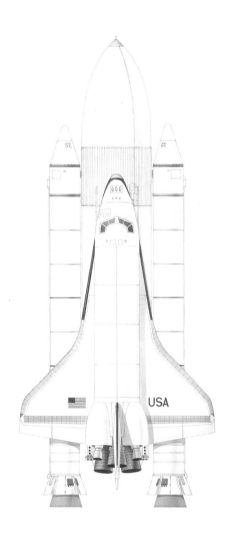

THE HISTORY OF
SPACE VEHICLES

TIM FURNISS

THUNDER BAY
P · R · E · S · S

This edition published in the United States by
Thunder Bay Press
an imprint of Advantage Publishers Group
5880 Oberlin Drive, San Diego, CA 92121–4794
www.advantagebooksonline.com

Library of Congress Cataloging-in-Publication Data available upon request

ISBN 1-57145-267-2

Editorial and design by
Amber Books Ltd
Bradley's Close
74–77 White Lion Street
London N1 9PF

Project Editor: Naomi Waters
Designer: Mark Batley
Picture research: Lisa Wren

Printed in Italy

CONTENTS

THE BEGINNING

Mankind's dreams of space exploration took a giant leap at Auburn in Massachusetts, USA on 16 March 1926. American rocket scientist Robert Goddard launched the world's first liquid-fuelled rocket to a height of 56m (183ft) in 2.5 seconds at 100km/h (62mph), becoming one of the first pioneers of space.

Although we may not view it as a remarkable achievement today, the use of a rocket with liquid propellant rather than crude solid propellant, such as gunpowder, was a major step towards attaining the sophistication and controllability required by rockets to explore space. Goddard's achievement was the pinnacle of his work, which began in 1903 with the investigation of rocket performance. He made several proposals, including the general theory of liquid-propellant, hydrogen-oxygen rocketry; the use of "step," or multiple-stage, rockets; the idea of sending a camera around distant planets and returning it to Earth; and even anticipating the use of ion propulsion. His detailed work was carried out between 1912 and 1916, and was reported in his famous publication, *A Method of Reaching Extreme Altitudes*, published in 1920 by the

Left: The first successful launch of a V2 missile from Peenemünde was made on 3 October 1942. The rocket reached an altitude of 192km (119 miles). Over 2700 V2s were launched during World War II.

"The rocket, in principle, is ideally suited for reaching high altitudes, in that it … does not depend upon the presence of air for propulsion."

Robert Goddard, *A Method of Reaching Extreme Altitudes*, 1920

Right: The world's first liquid propellant rocket was launched from Auburn, Massachusetts by American Robert Goddard on 16 March 1926. It reached an altitude of 56m (183ft) and a speed of 100km/h (62mph).

Smithsonian Institute. The paper confirmed Goddard's unique understanding of the nature of modern rocketry and spaceflight. "The rocket, in principle, is ideally suited for reaching high altitudes, in that it...does not depend upon the presence of air for propulsion," he wrote. Goddard was granted a number of patents, one of which covered his principle of "feeding of successive portions of propellant...into a combustion chamber, giving a steady propulsive force."

Goddard experimented with smokeless powder propellant but soon turned to liquid, writing in a Smithsonian report that "the advantages of the liquid propellant rocket are that the propellant materials possess several times the energy of powders, per unit of mass." In 1923, Goddard successfully fired a pump-fed, liquid oxygen gasoline motor in a test frame, preparing for his historic rocket launch three years later. His work progressed rapidly. In 1929, he launched a rocket equipped with a camera and instruments, and supported by funds from the Guggenheim Carnegie Institution, he moved to Roswell in New Mexico in 1930 to continue his test work.

His first Roswell-based rocket, 3.35m (11ft) long, was powered by a liquid oxygen and gasoline engine to a speed of 800km/h (497mph) and an altitude of 609m (1998ft). Goddard increased this altitude to a record 2.28km (1.42 miles) while he attempted to master the principles of control and stabilisation. One of the most significant aspects of Goddard's work was his rocket control, which he demonstrated further by test-firing a rocket guided by a revolutionary gyroscope, keeping it on a steady and predetermined flight path. This first stabilized rocket was launched on 19 April 1932.

From 1930 to 1935, Goddard experimented with different methods of stabilisation using pendulums and gyroscopically-controlled vanes, of which the latter was more successful. At the time, the US Government was more concerned with atom bombs and conventional weapons, and completely ignored Goddard's work. Remarkable progress was also being made elsewhere.

THE BIRTH OF THE V2 ROCKET

In other parts of the world, more seeds of the Space Age were being sown. In Breslaw, Germany, on 5 June 1927, Hermann Oberth established the Society for Space Travel (Verein für Raumschiffahrt, or VfR). A year later, the VfR successfully test-fired a rocket powered by liquid oxygen and kerosene, and later fired small rockets,

called Mirak and Repulsor, from a rocket-flying field in Raketenflugplatz, a suburb of Berlin.

In 1932 a demonstration flight of Repulsor was made for the German Army at Kummersdorf, 100km (62 miles) south of Berlin. The following year, when Adolf Hitler came to power, the Gestapo seized the Society's records and equipment, and the inevitable militarisation of the VfR's work followed. Financial support for Oberth's pioneering work dwindled in the late 1930s, partly due to the German Government's anxiety about his links with other rocket-launching groups abroad. After initially being unimpressed by rocket developments, Nazi leader Adolf Hitler became interested when he realized their potential for military use. He increased the Oberth group's annual budget of 80,000DM

Below: Robert Goddard is considered the pioneer of modern rocketry, having a unique grasp of technology which enabled him to devise the controlled delivery of liquid propellants into a combustion chamber.

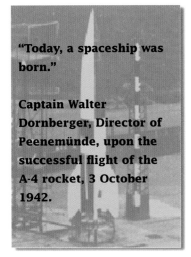

"Today, a spaceship was born."

Captain Walter Dornberger, Director of Peenemünde, upon the successful flight of the A-4 rocket, 3 October 1942.

to more than 11 million. The group then worked hard to overcome three major hurdles in rocket technology: steady operation of the motor, cooling of the components, and stabilisation.

A special section of the Army Weapons department was established at Kummersdorf under Captain Walter Dornberger, with the assistance of Wernher von Braun who had joined the Society in 1930 as an 18-year-old student. The launch of their first rocket, called the A-1 (Aggregate 1), was a failure. The A-1 was powered by a liquid oxygen and alcohol engine with a 300kg (661lb) thrust. The rocket was redesigned by von Braun and called the A-2. Much to the pleasure of the engineers, the A-2 tested successfully in 1934, reaching a modest 2.5km (1.6 miles) in altitude.

Encouraged by this, the Army pumped more money into rocket research, resulting in the development of the more powerful A-3 which contained many improvements, including a three-axis-controlled stabilization system using exhaust vanes and fin-mounted rudders. Increased funds flowed into the rocket group and further developments ensued. The massive scale

of this investment saw the building of a rocket launch base at Peenemünde in 1937 on the Baltic Coast and a factory at nearby Nordhausen, employing 12,000 people. The first fully controlled A-5 rocket was launched successfully from Peenemünde in 1939. Research then focused on the development of a "long-distance" rocket with a range of around 640km (398 miles), with the capability of accurately delivering a payload weighing 749kg (1651lb) to a target. The rocket was known as the A-4. It eventually became better known as the V2, one of the most destructive and deadly weapons of World War II, although the rocket entered the conflict almost at its end.

The five-and-a-half-tonne A-4 rocket operated with an engine thrust of 25 tonnes and was capable of firing for 68 seconds. It was powered by a liquid oxygen and alcohol engine, cooled by its liquid propellants before they entered the combustion chamber and, like Goddard's rockets, stabilized by graphite vanes in the exhaust. The booster flew successfully at the third attempt on 3 October 1942, reaching a height of more than 85km (53 miles), flying almost 200km (124 miles)

GERMAN V2 MILESTONES

Date	Development
13 June 1942	First test of A-4 vehicle from Peenemünde unsuccessful.
3 October 1942	First successful launch of V2 which travels 192km (119 miles).
17 February 1943	10th V2 rocket travels 193km (120 miles).
May–June 1943	More than 100 V2s test-fired from Blizna, Poland, launching 10 on one most of which are unsuccessful.
8 September 1944	First V2 to be fired in combat explodes in a Paris suburb; another strikes London a few hours later.
8 May 1945	At the end of the war, it is estimated that 1115 V2s have been successfully fired against England and 1675 against targets on the Continent.
December 1945	German V2 rocket engineers arrive in the US as part of Operation Paperclip. Others go to the Soviet Union.

Left: The first V2s launched in anger, on 8 September 1944, were aimed at Paris and London. One landed in Chiswick in the British capital, killing three people and injuring 10 others.

A9 POWERPLANT

The A9 missile, a prototype of a winged missile was launched in 1945, reaching an altitude of 90km and a speed of 4320km/h (268mph). It was based on V2 technology but had an improved combustion chamber.

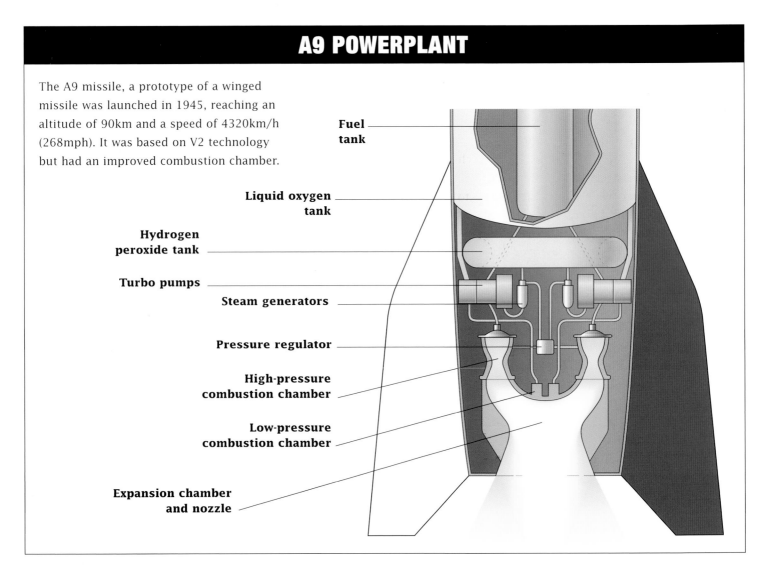

Fuel tank

Liquid oxygen tank

Hydrogen peroxide tank

Turbo pumps

Steam generators

Pressure regulator

High-pressure combustion chamber

Low-pressure combustion chamber

Expansion chamber and nozzle

"downrange" from Peenemünde. Captain Walter Dornberger, Peenemünde's chief, said, "Today, a spaceship was born." German engineers had already built a two-stage rocket and were making giant strides into space, with realistic hopes of launching rockets around the Earth, and even to the Moon. However, these aims were put to one side while a more urgent requirement existed – war.

THE VENGEANCE WEAPON

The V2 was to be equipped, not with a science payload, but with amatol, an explosive. On 8 September 1944, the first V2 – vengeance weapon – was launched on its journey from Peenemünde to Paris. Another V2, launched on the same date, hit

Chiswick in London at 6.44 p.m. with "a sound like a clap of thunder," killing three people and injuring 10 others. Sixteen seconds later, another V2 fell near Epping Forest, England, demolishing a number of wooden huts. Over the next 10 days, rockets continued to fall in and around London at a rate of about two a day, and by 17 September, 26 had fallen. A further 2789 V2s are estimated to have been fired at Britain and the Continent before the end of the war.

Meanwhile, Peenemünde's engineers had been developing an even more powerful rocket than the V2 – the prototype of the first intercontinental ballistic missile (ICBM). If it had been developed, this missile could have transformed Germany's

V2 ROCKET

The 25-tonne-thrust engine of the V2 fired for 63 seconds. The booster was powered by liquid oxygen and alcohol propellants, and carried a one-tonne explosive warhead. It first flew successfully in October 1942 as the A4 research rocket. "Today, a spaceship was born," said a project leader after seeing the missile rise up to 85km (53 miles) into the sky.

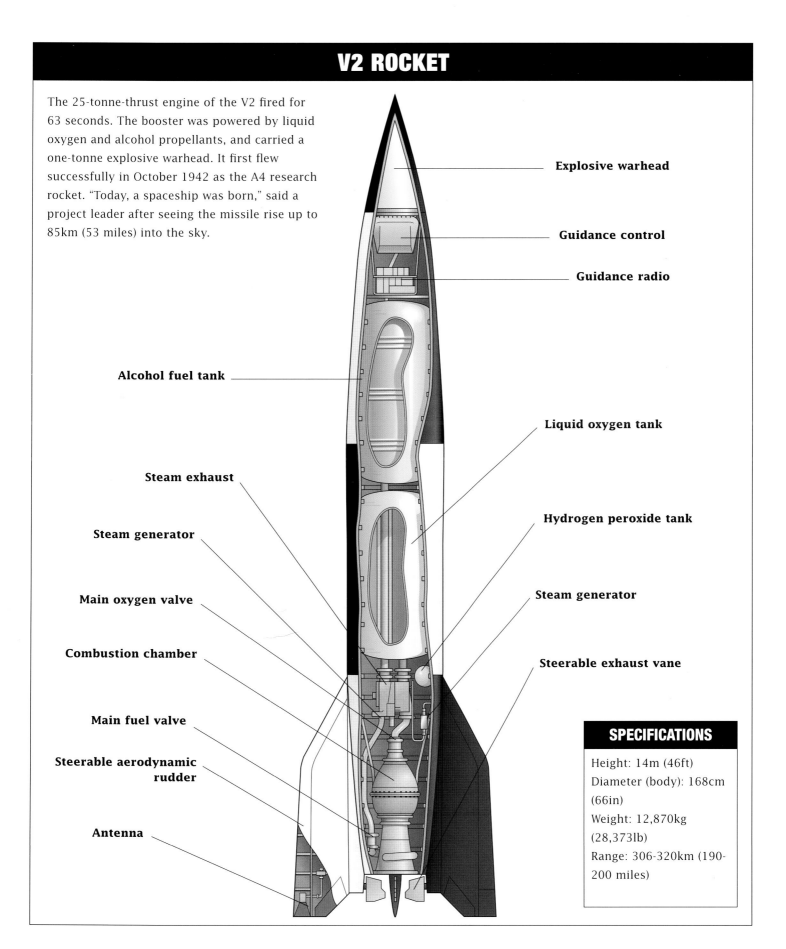

Explosive warhead

Guidance control

Guidance radio

Alcohol fuel tank

Liquid oxygen tank

Steam exhaust

Hydrogen peroxide tank

Steam generator

Steam generator

Main oxygen valve

Combustion chamber

Steerable exhaust vane

Main fuel valve

Steerable aerodynamic rudder

Antenna

SPECIFICATIONS

Height: 14m (46ft)
Diameter (body): 168cm (66in)
Weight: 12,870kg (28,373lb)
Range: 306-320km (190-200 miles)

WINGED V2 ROCKET

German rocket scientists hoped one day to develop a winged, multi-stage rocket that would be able to place a satellite into orbit, but military applications took priority, and the A9 prototype was seen as a precursor to an intercontinental ballistic missile.

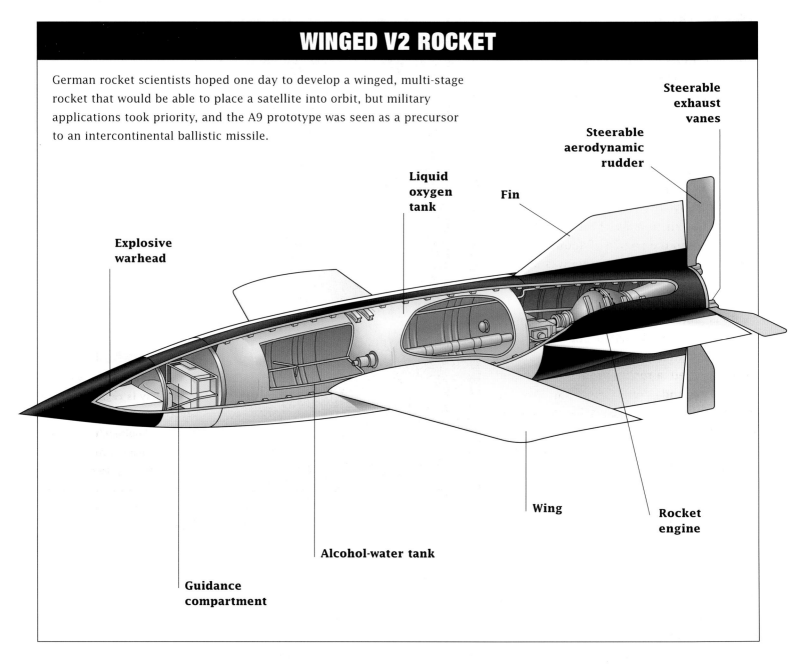

Steerable exhaust vanes

Steerable aerodynamic rudder

Fin

Liquid oxygen tank

Explosive warhead

Wing

Rocket engine

Alcohol-water tank

Guidance compartment

fortunes in the war by being able to drop a bomb on the US. On 24 January 1945, the A-9 winged prototype was successfully test-fired, reaching an altitude of about 90km (56 miles) at a maximum speed of 4320km/h (2684mph).

Despite their military work, von Braun and his engineers still had dreams of developing a winged rocket plane which would be launched piggyback-style on a recoverable booster rocket – not unlike today's Space Shuttle. Other plans included a multi-stage piloted rocket designed to reach a speed of 28,800km/h (17,895mph) which, as von Braun put it, "would not return to Earth, as gravity and centrifugal force would balance each other." In other words, this rocket would orbit the Earth. If World War II had not intervened, Germany could have become the pioneering country of the Space Age, which was also taking root further east. In addition to the work of American and German engineers, Russia was making parallel strides towards space, but their work was little publicized.

TSIOLKOVSKY –
THE "FATHER OF SPACE FLIGHT"

Russia's Konstantin Tsiolkovsky (1857–1935) is regarded as the "father of space flight." His theoretical works include the practical theory of using liquid-propellant controlled rockets. Tsiolkovsky was a humble schoolteacher, but in 1883 he established that a rocket could operate in the vacuum of space, not because it is propelled by the force of the exhaust against the air, but by the force of being expelled from the nozzle of another rocket. Although obvious now, this was a vital statement at the time. That was not the end, for Tsiolkovsky continued to write copious notes and theories on advanced concepts of liquid-fuelled spaceships and the mathematics of space travel. In 1903 he even designed a rocket propelled by liquid hydrogen and liquid oxygen "as an explosive mix of hot gases being expelled from a nozzle."

Tsiolkovsky continued his work, and considered other types of propellants. He proposed the use of a revolving flywheel – a gyroscope – to stabilize rockets in flight, and the use of vanes in the exhaust outlet to steer the rocket. Another innovation was his outlining of a multi-stage vehicle, and he showed how a series of rockets could fly in a "step" principle. As stages became depleted of their propellants, they dropped off, so reducing the weight of the main vehicle. Without carrying any excess weight, the vehicle would be able to travel fast enough to reach orbit. The brilliant professor surmised that when the speed of a rocket reached 8km (5 miles) per second, the centrifugal force would overcome the gravitational force, which drags projectiles back to Earth, and they would enter a continuous arc around Earth – or an orbit.

Some of Tsiolkovsky's early rocket theories were put into practice by Soviet engineers. Six years after the founding of the Soviet Union in 1922, the Leningrad Gas Dynamics Laboratory (GDL) was formed to

Left: Russia's Konstantin Tsiolkovsky is regarded as the 'father of spaceflight.' The humble schoolteacher's early rocket theories were put into practice, leading to the development of Russia's first liquid-propellant rocket engine.

develop more advanced versions of their existing military solid rockets. The result was the evolution of liquid-propellant and monopropellant rockets. The first liquid-propellant rocket engine, called ORM 1, was static-tested in 1931. After more than 40 tests, work began on a series of further ORM engines, vital engine components, such as combustion chambers, and the use of different propellants, including gasoline and nitric acid, were developed.

At the same time, the Group for the Study of Reaction Propulsion (GIRD) had also been established in 1931 to work primarily on the development of liquid-propellant engines. The following year, the Soviet Government established a branch of GIRD, called the Rocket Research and Development Centre in Moscow. Its director

SERGEI KOROLEV

Although the name Korolev had been heard of before World War II when news about Soviet rocket engine achievements had become known, he died unknown by name. On 14 January 1966, Korolev died after an operation and was revealed as the person who had, for ten years, been known simply as the "Chief Designer", the man who launched the ICBM, Sputnik, and Lunik, and who spoke to Yuri Gagarin, the first man ever to travel into space from the launch blockhouse in 1961. A hero of the Space Age remained nameless – until after he had died.

Korolev was born in the Ukraine on 30 December 1907, and worked as a test pilot for the aircraft industry and a designer of early Soviet planes and gliders. In 1932, he took over the management of GIRD's design and production division, the group responsible for developing liquid-propellant engines, and later to become part of RNII (Scientific Rocket Research Institute). Korolev wrote a book *Rocket Flight in the Stratosphere* in 1934, and then converted theory into practice by helping to develop rocket-powered aircraft and gliders. He was later responsible for developing advanced versions of the V2.

Korolev designed the ICBM, for which approval was granted in 1954, and worked in the desolate steppes of Kazakhstan near a railway junction called Tyuratam. Living in tents and huts, and working in extreme conditions of cold in winter and heat in summer, Korolev began the development of what became known as the Baikonur Cosmodrome, Russia's equivalent of Cape Canaveral. After the ICBM launch, Korolev was responsible for the launch of Sputnik 1 and for the development of almost all the new rockets and spacecraft, manned and unmanned, that were used to reach some of the great milestones in space history. Korolev was also the "father" of the Soviet cosmonaut team of young test pilots who made the first space flights.

Above: Sergei Korolev helped establish the Rocket Research and Development Centre (GIRD) in 1932, and later led the development of the world's first intercontinental ballistic missile (ICBM) at a secret base near Tyuratam, Kazakhstan.

was an engineer and pilot called Sergei Korolev. The Korolev team first developed a series of liquid-propellant rockets which used liquid oxygen and a solid gasoline, rather like a gel. A rocket called the GIRD 9 was launched on 17 August 1933 and reached an altitude of about 1.5km (0.93 miles), eventually becoming the Soviet Union's first anti-aircraft rocket. The second rocket, called the GIRD 10, was the first fully liquid-propellant rocket developed by the Soviet Union and was launched on 25 November 1933, when it reached an altitude of approximately 4.9km (3 miles).

The Kremlin soon realized the potential of this rocket work for military purposes, and combined the GDL and GIRD into the Scientific Rocket Research Institute (RNII). In 1939, the RNII successfully test-fired a two-stage hybrid rocket which reached an altitude of about 1.8km (1.12 miles). By this time, all work was classified, as World War II approached, and except for small military missiles, Soviet rocket development all but stopped during the conflict. In May 1945, the Soviet Union and the United States, destined to become space superpowers in the post-war era, were to be linked in an inexorable way – at Peenemünde.

THE "SPACE RACE" BEGINS

On 2 May 1945, Peenemünde's days as a rocket development centre were over as it was surrounded by the advancing US, British and Russian armies. The chance to obtain details of the rocket programme was not lost on either the west or the east. By various means, engineers, plans and left-over parts of rockets all found their way into the US and Russia, and provided a vital infusion of experience and knowledge for these countries' fledgling rocket programmes.

Von Braun and many of the Peenemünde team were taken to the US to work for the Army at the White Sands Missile Range in New Mexico, where a new version of the V2 was developed for scientific purposes. Many of the V2s were launched with science payloads for the study of the upper atmosphere. Eventually the V2 was given an upper stage, called the WAC Corporal, which itself had made several science flights into the upper atmosphere. Named "the Bumper," this new combination made deeper flights into space. In 1950, a launch of the rocket inaugurated operations at a remote alligator and mosquito-infested site on a Florida sand spit, whose name would become synonymous with the Space Age – Cape Canaveral.

Another US rocket, based on V2 technology, was the Aerobee, which also flew successfully into the upper atmosphere. At the same time, at White Sands, the US-developed Viking research rocket (originally called Neptune) was flown to similarly high altitudes of up to 252km (157 miles), as scientists gleaned more information about the upper atmosphere and the space environment. The Viking embodied many of the features that would be used on later military missiles, such as lightweight, integral tank structures, swivelling engines for steering, and opening nose cones to expose scientific instruments in space.

Meanwhile in May 1945, the Second White Russian Army under General Konstantin Rokossovsky had also occupied Peenemünde. There they had found a veritable treasure of V2 rocketry, and had rounded up specialists to take on a journey to the east, to some discomfort and obscurity compared with their colleagues heading west. In 1946–47, V2 production in the Soviet Union was greater than that ever achieved by Germany, and boosters were soon flying regularly into the Earth's atmosphere carrying science packages.

Below: The first Soviet rocket, GIRD 09, was launched on 17 August 1933, reaching an altitude of 1.5km (0.93 miles). It was followed by another model, GIRD 10, which reached 4.9km (3 miles) on 25 November 1933.

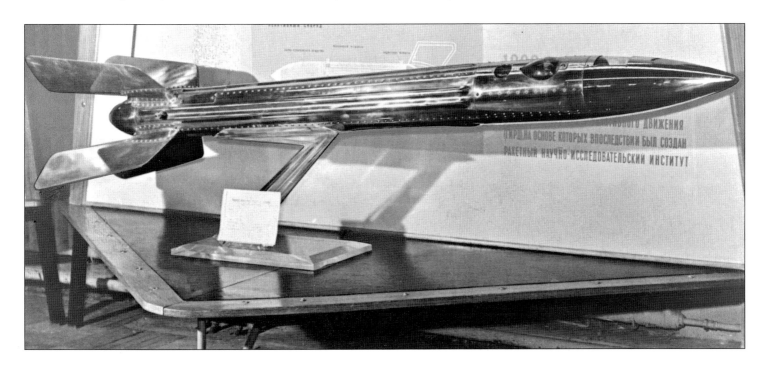

At the end of the war, von Braun, using a stolen train, led 500 people to surrender to a US army private. Von Braun and 126 fellow "Peenemünders" were subsequently stationed at White Sands, where they worked on US versions of the V2.

OPERATION PAPERCLIP

The US Operation Paperclip and its Soviet counterpart, which released Germany's advanced knowledge and experience in rocketry to both superpowers, were two pivotal events in the history of space exploration. As the German rocket launch base at Peenemünde on the Baltic coast was about to be surrounded by the Allies, and with the war clearly coming to an end, Wernher von Braun decided that if he and his team were going anywhere it would be to the US, and not eastwards with the Red Army.

The US Army arrived at the nearby Nordhausen factory to find an amazing horde of missiles. Soon Operation Paperclip was put into action to evacuate complete V2s,

Above: Wernher von Braun (right), his arm encased in plaster after a motorbike accident, surrendered to the 44th Division at the end of World War II, beginning his journey to the USA and Apollo.

along with the scientists, to White Sands Proving Range in New Mexico, USA. Von Braun and a few members of his team surrendered to the US 44th Division and 14 tonnes of V2 documents were recovered, together with about 100 vehicles. However, the US forces were not told that Nordhausen would lie within the Soviet zone at the end of the war, so no attempt was made to destroy the factory. The remaining V2s and documents eventually fell under Russian control, together with some of the engineers. Other V2 team members had fled. These "lost" scientists and technicians, and those who had already surrendered, were offered five-year contracts to work in the US. The first German V2 workers arrived in the US on 20 September 1945, and by 1954 there were 50 German engineers based in Alabama as naturalised Americans.

Under the leadership of Korolev and other former GIRD engineers, the Soviets soon developed their version of the V2 called the T-1. This version, in turn, was soon followed by a single-stage geophysical rocket able to fly to altitudes approaching 100km (62 miles), with a payload of 130kg (287lb). Another rocket, called the Mitio or T-2, increased this capability to 190km (118 miles) and 79kg (174lb) respectively. The T-2 was to be the workhorse of the proposed International Geophysical Year (IGY) of 1957–58, in which the Soviet Union had agreed to participate, as had the Americans with their equivalent, the Viking. The T-2 was also the Soviet's Intermediate Range Ballistic Missile (IRBM), and the basis of the ICBM.

By 1957, a new rocket, the T-3, had been developed and rocket technology made a giant leap forward. It reached an altitude of 211km (131 miles) with a payload weighing over two tonnes! This rocket actually became the world's first ICBM, for once again military matters took precedence. With the onset of the Cold War, both the west and east possessed the proven technology to produce nuclear weapons, versions of which so effectively ended World War II at Hiroshima and Nagasaki. Delivery of the weapons into enemy territory could only be accomplished by long-range aircraft, or missiles with intermediate or intercontinental range. Both the US and Soviet Union continued to develop their space science programmes, with some research rockets so advanced that they carried recoverable nose cones containing science instruments and even animals. However, there was also a need to develop long-range missiles.

At first, the US relied upon the capability of Hustler bombers being able to fly into Russia to deliver nuclear bombs if necessary. At the same time, Russia proceeded with its development of an ICBM. The concept of the Soviet Union's ICBM had been born as early as 1946, when the Soviet Air Force quickly realized that in a potential Soviet-American conflict they could not only rely on short-range V2s, and that long-range rockets would be essential. The development of an IRBM was approved in 1949, and the ICBM in 1954. By the time the US realized that it needed an ICBM, Russia was well on the way to test-firing its first missile. As a step towards the development of the ICBM, the US started to build the Redstone, an IRBM, using von Braun's V2 technology. The US left it until

WERNHER VON BRAUN

The man who was to mastermind the development of the Saturn V rocket which sent the first men to the Moon was born in Wirsitz in Germany on 23 March 1912, the second of three sons to Baron Magnus von Braun. In 1932, Wernher von Braun received a degree in mechanical engineering and was offered a grant to conduct scientific investigations into rocket engines. Von Braun joined the Verein für Raumschiffahrt rocket society and developed a rocket with $400 funding from the German Army. He impressed Army Captain Walter Dornberger, and the two men, assisted by a team of 80 engineers, established a rocket centre at Kummersdorf in 1934, from where two rockets were successfully developed and launched. Later the team moved to Peenemünde to develop the V2.

At the end of the war, von Braun, using a stolen train, led 500 people to surrender to a US private. The German SS had been issued with orders to kill the V2 rocket team. Von Braun and 126 fellow "Peenemünders" were subsequently stationed in Fort Bliss, Texas, USA, and worked at White Sands on US versions of the V2. In 1950 they were transferred to Huntsville in Alabama, where they set up the Redstone Arsenal, developing the Redstone Intermediate Range Ballistic Missile (IRBM). Von Braun's team then developed a version of the Redstone as a satellite launcher, called the Jupiter C, which eventually launched the US's first satellite, Explorer 1, in 1958. Two years later, the newly formed National Aeronautics and Space Administration (NASA) established the Marshall Space Flight Centre in Hunstville and von Braun became its first director.

Marshall's main programme was the development of the Saturn boosters for the Apollo lunar programme, and in 1970 he was appointed deputy associate administrator for planning at NASA HQ in Washington DC. But von Braun resigned from NASA disappointed that the space agency's vision of the future of space differed from his own, and he became vice-president of Fairchild Industries in 1972. He was later diagnosed as suffering from cancer, retired from Fairchild in 1976 and died on 16 June 1977.

Above: German-born Wernher von Braun led the development of the Saturn 5 booster which launched the first men to land on the moon in 1969. He died of cancer in 1977.

THE V2 IN THE US

Date	Development
16 January 1946	The US upper-atmosphere research programme begin using captured German V2 rockets, of which more than 60 are eventually launched before US versions are built. Using V2 technology, the Applied Physics Laboratory develop the Aerobee medium-altitude rocket, while the Naval Research Laboratory develop the Neptune high-altitude rocket, later known as Viking.
15 March 1946	First US-assembled V2 is static tested at White Sands.
16 April 1946	First US-assembled V2 is launched, and in July flights 5 and 9 set new altitude records of over 180km (111 miles), with flight 17 reaching a speed of 5760km/hr (3579mph).
24 October 1946	V2 flight 13 takes motion pictures of Earth from 104km (65 miles) altitude.
17 December 1946	V2 achieves 185km (115 miles) altitude record.
20 February 1947	First in a series of V2 firings, the 20th, known as the Blossom Project, tests the ejection of a canister and its recovery by parachute, containing fruit flies and various types of seeds exposed to cosmic rays.
7 March 1947	US Navy V2 take the first Earth picture from 180km (111 miles) distance.
6 September 1947	V2 is launched from the US aircraft carrier *Midway* but explodes after a 9.6km (6 mile) flight.
6 February 1948	V2 flight with electronic flight-control unit tests to 112km (70 miles).
13 May 1948	V2 with US Wac Corporal second stage (Bumper Wac), flies to 126km (78 miles) from White Sands.
30 September 1948	Bumper Wac 3 reaches 149km (93miles).
24 February 1949	Bumper Wac reaches record 390km (242 miles) and 8240 km/hr (5120mph) from White Sands.
14 June 1949	V2 carries monkey, Albert II, to 132km (82 miles) but animal died on impact.
24 July 1950	Bumper Wac 8 flies on first launch from Cape Canaveral.
29 July 1950	Bumper Wac 7 launches from Cape Canaveral and attains a speed of Mach 9 (nine times the speed of sound), the fastest yet achieved.
29 October 1951	V2 number 66 concludes the use of German missiles by the US V2 launch summary 66 launches, of which eight are Bumper flights; two are launched from Cape Canaveral and 64 from White Sands.

1954 before starting a crash programme to develop the Atlas ICBM. It did not make its first flight until 17 December 1957, powering its way out of Cape Canaveral with a thrust of 176.5 tonnes.

AIMING FOR SPACE

By 1955, sufficient technology existed for the US to announce that a science research rocket, called Vanguard, would launch a research satellite in 1957 which would make observations as part of the IGY. The rocket was based on the Viking sounding rocket which had made 12 research sub-orbital flights from White Sands. Russia also announced that it would launch a satellite but this was ignored, the impression being that Russia was far less technically capable than the US, mainly because of the secrecy surrounding its work. The Vanguard programme was to be a civilian affair, but managed by the US

Above: The first US-assembled V2 was launched in April 1946 from White Sands, New Mexico. Later versions were equipped with the US Corporal missile as a second stage. The new vehicle was called Bumper Wac and first flew in 1948.

Left: Two Bumper Wac V2-Corporal missiles were launched from Cape Canaveral, Florida, a new rocket-proving range on the Atlantic coast. The first launch from Canaveral was of Bumper Wac 8 on 24 July 1950.

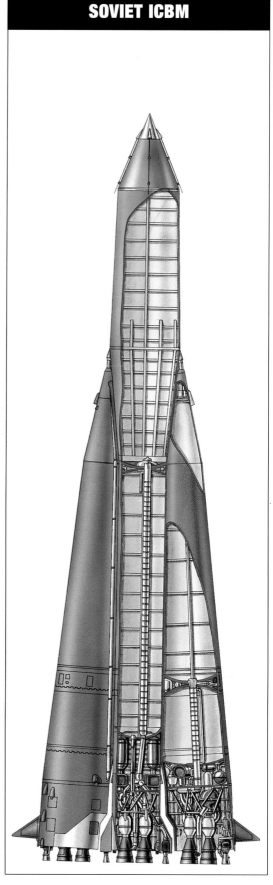

SOVIET ICBM

Above: Twin Aerobee sounding rockets are launched from White Sands, New Mexico.

Right: The first ICBM was developed by a Soviet team, and designated the R-7. It was the forerunner to the launch vehicle which carried the first satellite into orbit.

Navy which had been chosen to develop the IGY rocket after a competition with the Army and the Air Force. The Air Force proposed a version of the winged ballistic missile Bomarc with a piggyback stage, while the US Army team, led by von Braun, proposed a version of the Redstone, called the Jupiter C, both designed for military purposes. Wary of adverse publicity, President Dwight D. Eisenhower did not wish to have a military-based missile

VIKING ROCKET FLIGHT LOG

Date	Flight Log
3 May 1949	Flight 1 reaches 80km (50 miles), impacting at T+291 seconds after premature engine cut-off due to steam leaks in the engine turbine.
6 September 1949	Flight 2 reaches 51km (32 miles), impacting at T+394 seconds after similar leaks.
9 February 1950	Flight 3 reaches 80km (50 miles), impacting at T+420 seconds after cut-off due to deviation from flight path.
11 May 1950	Flight 4 reaches 168km (104 miles), impacting at T+435 seconds after the first and only launch from a ship, the USS Norton Sound.
21 November 1950	Flight 5 reaches 172km (107 miles), impacting at T+450 seconds after a longer burn time due to reduced thrust.
11 December 1950	Flight 6 reaches 64km (40 miles), impacting at T+292 seconds after a night-firing in which the stabilising fins fail, causing the rocket to perform violent manoeuvres.
7 August 1951	Flight 7 reaches 196km (122 miles), impacting at an estimated T+530 seconds on the highest flight with the original air frame. Flight records the highest altitudes for measurements of atmospheric winds and density.
6 June 1952	Flight 8 reaches 6.4km (4 miles), impacting at an estimated T+100 seconds after the rocket breaks loose during a static test-firing and self-destructs.
15 December 1952	Flight 9 reaches 216km (134 miles), impacting at an estimated T+540 seconds in the first successful flight of the new airframe.
7 May 1954	Flight 10 reaches 196km (122 miles) in 290 seconds, impacting at T+295 seconds after the motor explodes in the first attempt; the rocket is rebuilt. Flight makes first measurements of positive ion composition at high altitude.
24 May 1954	Flight 11 reaches 252km (157 miles), impacting at T+557 seconds on the highest Viking flight. Flight also takes the highest altitude images of the Earth to date.
4 February 1955	Flight 12 reaches 230km (143 miles), impacting at T+540 seconds on the final successful flight of the programme.

Right: Twelve Viking missions were flown between 1949 and 1955, reaching a maximum altitude of 252km (157 miles). The Viking was developed by the Naval Research Laboratory and the Martin company.

VIKING

The Viking rocket was developed to reach altitudes of over 240km (149 miles), and as a precursor to a satellite launcher. It was launched from White Sands from the same gantry used to launch the V2.

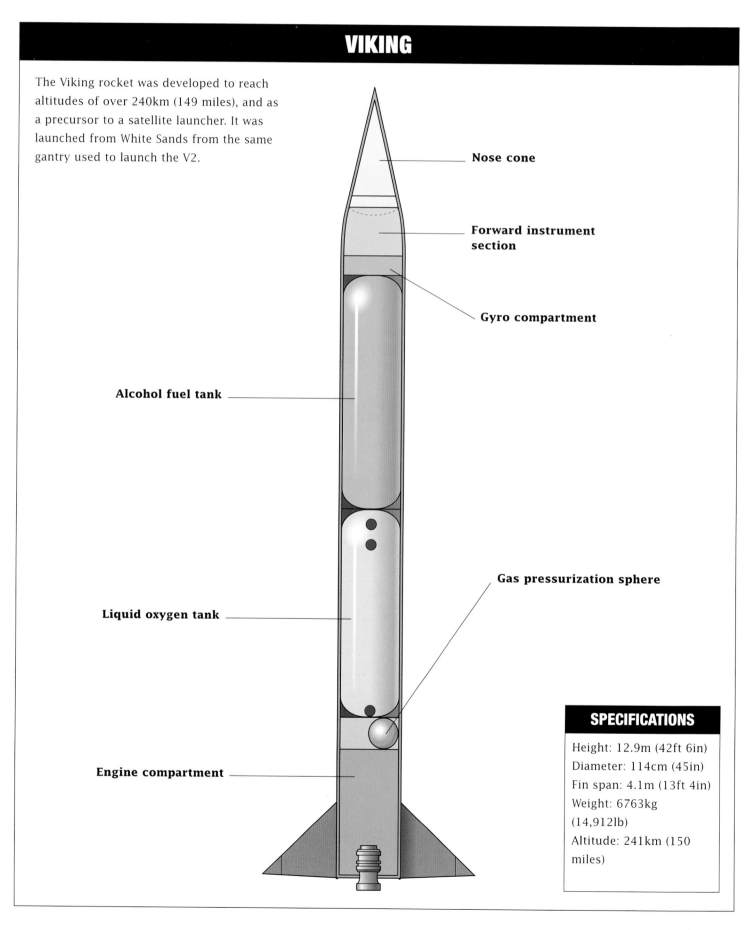

Nose cone

Forward instrument section

Gyro compartment

Alcohol fuel tank

Gas pressurization sphere

Liquid oxygen tank

Engine compartment

SPECIFICATIONS

Height: 12.9m (42ft 6in)
Diameter: 114cm (45in)
Fin span: 4.1m (13ft 4in)
Weight: 6763kg (14,912lb)
Altitude: 241km (150 miles)

launching a US science satellite. Little did he know that Russia was planning to do exactly that.

Vanguard secured the go-ahead while a frustrated von Braun continued to develop the Jupiter rocket for future scientific applications. On 7 August 1957, von Braun launched a Jupiter C to an altitude of 960km (597 miles). If it had been equipped with an upper stage, it would have placed the first satellite in orbit, but Eisenhower had forbidden von Braun to launch the first satellite. The Vanguard, meanwhile, was experiencing develop-ment problems and it did not appear likely that the US's satellite would be in orbit until 1958.

At the same time, in a remote rocket base at Tyuratam in Kazakhstan, the Soviet rocket engineer Korolev was preparing the first ICBM for launch. On 26 August, the Soviet Union announced that on 3 August 1957 it had launched the "first super long-distance, inter-continental, multi-stage ballistic rocket." The rocket was based on the concept of clustering five former geophysical rockets together as a single stage. Four of these geophysical rockets, "strapped-on" to the fifth, would be jettisoned at high altitude, leaving a single booster to continue on its journey. With the ICBM objective achieved, Korolev was soon preparing the next ICBM for something much more spectacular.

Above: The US Army's Redstone IRBM was a direct descendant of the V2. Launches of a prototype vehicle, code-named Hermes, were made from White Sands.

Left: Von Braun (second from right) established the Redstone Arsenal in Huntsville, Alabama in 1950, and developed the first Redstone Intermediate Range Ballistic Missile (IRBM). The Redstone was the direct ancestor of the Saturn V Moon rocket.

THE FIRST FIVE YEARS

The Soviet Union was a republic very conscious of its anniversaries. It had planned to launch the world's first satellite on 14 September 1957 on what would have been Tsiolkovsky's 100th birthday. But the launch of the modified R7 missile from Tyuratam in Kazakhstan was delayed until 4 October, which then became more than just a Soviet anniversary – it went down in history as the day that the Space Age truly began.

K orolev's booster thundered into the night sky at 22:28 hours and 4 seconds, Moscow time. Around five minutes later, at an altitude of about 215km (134 miles), the core stage of the booster reached a velocity sufficient to orbit earth for the first time in history, travelling at a speed of 7.99km per second (4.96 miles per second), making a continuous arc around Earth without falling back under the influence of the planet's gravity. The arc (or orbit), which in this case crossed the equator at an angle, called the

Left: US Naval Research Laboratory scientists discuss the solar cells on the Solar Radiation 3 satellite, Solrad, which was launched in 1961.

THE SPUTNIK EFFECT

Western assumptions that the Soviet Union was a backward nation, despite producing the atomic bomb, were exposed as folly when the Sputnik 1 satellite appeared in space. The Soviets had beaten the US in the race into space – and proved that they had a rocket that was capable of carrying a nuclear warhead into the US: an Intercontinental Ballistic Missile (ICBM). At a time when the Cold War was at its most intense, and some Americans were building bunkers in their back yards, Sputnik was a frightening event. Rather than celebrating the birth of the Space Age, the west felt threatened. It seemed imperative to get a US satellite into space as soon as possible. However, its satellites were nothing to match the size of Sputnik, and the rockets were less powerful, too. America had not yet declared its ICBM, the Atlas, operational.

President Dwight D. Eisenhower ordered the US Navy to launch its Vanguard booster as quickly as possible and to get something into space. So, what was intended to have been a mere test flight of the Vanguard rocket, with a tiny test satellite which may or may not have reached orbit, became a well publicized satellite launch attempt. The event was watched on live TV in the US (Soviet launches happened in secret) as the rocket rose a foot from the launch pad at Cape Canaveral but lost thrust. It fell back and exploded, tipping the test satellite off the top the rocket. The press had a field day: "Kaputnik" and "Flopnik" were the best of the headlines, and much hand-wringing occurred.

Eisenhower called into service von Braun's Jupiter C. The Vanguard was considered a civilian venture because the launcher was not a military rocket, and had always been prioritized over von Braun's Jupiter, basically an uprated Redstone Intermediate Range Ballistic Missile (IRBM). Eisenhower was keen for the US satellite to be a scientific and non-military venture. After the advent of Sputnik, that policy was abandoned and the Space Age was never going to be quite the same. It soon became known as the Space Race, with the US and Soviet Union competing against each other. Each flight was compared to assess respective national leaderships, and the Soviets always seemed to be the nation which achieved the "space firsts".

inclination, of 65.1°. A small nose cone was ejected from the core stage. An even smaller nose cone was then released exposing a 58cm (23in) diameter silver sphere, which in turn was spring-released, from its carrier. Four previously deployed aerials, approximately 3m (9.8ft) in length, sprang into position ready to transmit. Later, the 82kg (181lb) satellite reached an orbital high point, the apogee, of 939km (583 miles). The lowest point of the orbit, the perigee, was the approximate point at which the satellite was ejected, and the time it took for the satellite to make one orbit of the Earth – one hour 36 minutes – was called the period.

Radio Moscow announced the news to a stunned world a little later on 5 October. The satellite was called Sputnik, the Russian word for "traveller". The spacecraft was equipped very basically, with two radio transmitters weighing 3.5kg (7.7lb) and which operated on frequencies of 20 and 40 megacycles. The signals were picked up by radio receivers all over the world, including the west, and their "bleep, bleep" sounds were soon to become synonymous with the dawning of the Space Age. The sounds also came to haunt the western world, particularly the US, which had been beaten into space by a nation assumed to be backward, and without the technology to match America's might.

In a minor way, Sputnik demonstrated one of the first applications of space satellites. Temperature sensors located inside the satellite sphere, pressurized by nitrogen gas, caused changes in the frequency of the telemetry signals. These signals enabled engineers to monitor the effects of temperature variations in the craft. The propagation of the signals in the Earth's atmosphere produced data on the ionosphere, while tracking the orbit of the satellite and its carrier stage provided data on how the orbit was being affected by atmospheric drag. The tracking of Sputnik's signals by the west was the first step towards establishing today's Global Positioning System (GPS) navigation satellites. The exact orbital position of Sputnik was determined by the Doppler shift of its signals. The so-called Doppler effect is the change in the apparent frequency of sound, light or other wavelengths resulting from the motion between their source and an observer, enabling the observer to ascertain their position in relation to the object.

According to international rules, the objects in orbit were designated 1957 Alpha 1, 2 and 3. The first object – 1957 Alpha 1 – was the spent-rocket stage, weighing around 7.5 tonnes, which somewhat humourously signified that the first designated object in space was also the first piece of space debris! Sputnik was designated 1957 Alpha 2, and the nose cone, another piece of debris, was named 1957 Alpha 3. Sputnik was battery-powered and, much to the relief of the west, its haunting "bleep, bleep" sounds ended 21 days later as the battery power finally drained away. The upper atmospheric drag gradually slowed down Sputnik, and it eventually entered the denser layers of the Earth's atmosphere. Still travelling at high speed, it was destroyed by friction on 4 January 1958. This course became known

as "re-entry". The booster and nose cone suffered a similar fate.

SECOND BLOW FOR AMERICA

Before the US had a chance to reply to the Sputnik, a second Soviet craft entered orbit on 3 November. Designated 1957 Beta, the craft was named Sputnik 2, weighing an extraordinary 508.3kg (1121lb). It also created even more consternation in the west because it carried the first living creature in orbit, a husky dog called Laika. Like the US, the Soviet Union had been sending animals into space – in the Soviets' case, mainly dogs – for many years on sub-orbital rocket research flights, but Laika was the first living creature to orbit. The conical-shaped 4m (13ft long) Sputnik 2 comprised a spherical container like Sputnik 1, and a cylindrical container in which Laika was housed. The craft remained attached throughout the flight to its seven-tonne-plus rocket stage in its orbit between its perigee of 212km (132 miles), and its apogee at 1660km (1037 miles), at an inclination of 65.3°.

Telemetry from the spacecraft provided the first information about solar and cosmic radiation in space, and particularly suggested the existence of a radiation belt surrounding Earth. Laika was destined to

Below: A technician prepares a 58cm (23in) diameter shiny sphere, with four antennae, for launch. The sphere would become known as Sputnik 1.

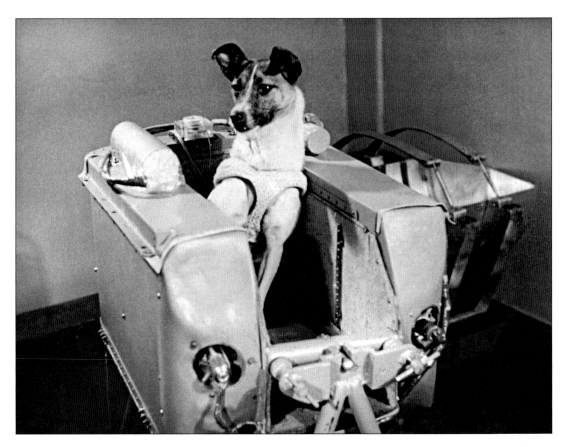

Right: Probably the most famous animal in the world. Laika is prepared for her flight into history – the first living creature to be placed into orbit. The air in Sputnik 2 ran out after a week, and Laika's inevitable death angered many animal lovers.

die in space when her air supply expired after a week. The little dog was provided with oxygen by a highly reactive chemical which also absorbed the carbon dioxide and water vapour released in the capsule. Sputnik re-entered earth's atmosphere on 14 April 1958.

With the Soviet lead in space increasing and the US left earth-bound, there was considerable political pressure to put a US satellite into space as quickly as possible. A tiny test-article satellite for the US Navy's Vanguard programme was called into service, weighing 1.35kg (2.9lb), the size of a grapefruit. The launch attempt on 6 December 1957 from Cape Canaveral, aboard the elegant and rather slender civilian Vanguard rocket, has gone into history as "Kaputnik". Following ignition, the Vanguard booster briefly rose a few feet before it fell back and exploded spectacularly. The US Air Force team, led by the frustrated Wernher von Braun was implored by President Eisenhower to launch an Explorer satellite on a military missile Jupiter C booster as quickly as possible.

THE US IN ORBIT

The launch on 31 January 1958 from Pad A of Complex 26 at Cape Canaveral was a success. The 13kg (28.7lb) pencil-shaped satellite entered an orbit of 33.3° inclination, with a perigee of 356km (221 miles) and an apogee of 2548km (1583 miles). Called Explorer 1, the US's first satellite was destined to make the first great space discovery. Explorer 1 was equipped with Geiger tubes to detect cosmic rays, along with external and internal temperature gauges, and micrometeorite detectors, including an ultrasound microphone, to detect impacts. The Geiger counters, provided by Dr James Van Allen of the State University of Iowa, went into action as the Explorer satellite approached the apogee of its orbit. Van Allen rightly concluded that the counters

Left: America's first attempt to
launch a satellite, Vanguard,
on 6 December 1957 fails at
lift-off at Cape Canaveral.
The Vanguard booster rose
a few feet, fell back and
exploded. The disaster became
known as "Kaputnik" to the
world's media.

EXPLORER

America's first satellite, Explorer 1, was launched by a Jupiter rocket on 31 January 1958, and discovered radiation belts encircling the Earth.

SPECIFICATIONS

Length: 205cm (80.7in)
Total weight: 14kg (31lb)
Instrumentation weight: 8.32kg (18.3lb)

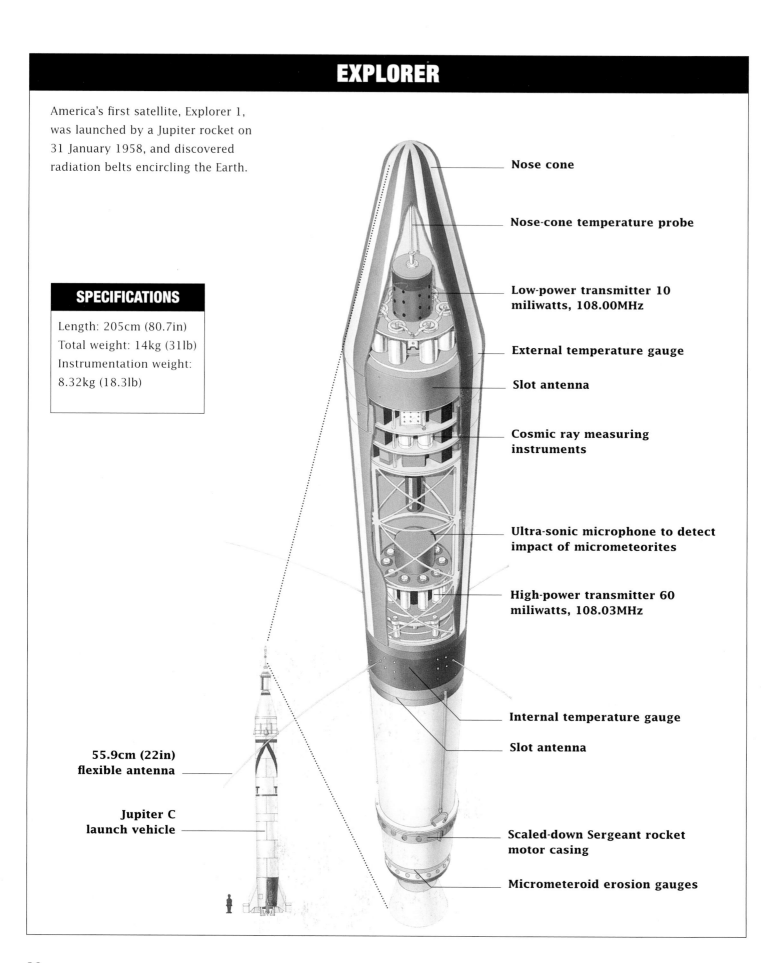

Nose cone

Nose-cone temperature probe

Low-power transmitter 10 miliwatts, 108.00MHz

External temperature gauge

Slot antenna

Cosmic ray measuring instruments

Ultra-sonic microphone to detect impact of micrometeorites

High-power transmitter 60 miliwatts, 108.03MHz

Internal temperature gauge

Slot antenna

Scaled-down Sergeant rocket motor casing

Micrometeroid erosion gauges

55.9cm (22in) flexible antenna

Jupiter C launch vehicle

SATELLITE SPOTTING

All over the world people were able to participate in the exciting early days of the Space Age able to watch satellites pass overhead, looking like stars moving across the sky a little faster than airliners moving at high altitude. The satellites could be seen because they reflected the sunlight shining on them as they orbited in space. As they passed into Earth's shadow, as seen from their orbit, their light faded and they went into eclipse. The large rocket body that launched Sputnik 1 was easily seen, as was Sputnik 2, which was still attached to its rocket body. Sputnik 1 could only be seen clearly using binoculars. The 30m (98ft) diameter Echo 1 balloon, which was covered with aluminium film was the brightest and the most watched, becoming a symbol of the

Above: A long exposure image reveals the Echo satellite moving across the night sky like a wandering star.

Space Age for people all over the world. Some satellites appeared to flash as they flew overhead. This was the result of their rotating or spinning and catching the sunlight intermittently. Long exposure photography captured some satellites passing overhead as a streak of light crossing the star-filled night sky.

Today Vanguard 1 is the oldest man-made object still in space. With its relatively high orbital period of two hours 13 minutes, the satellite has avoided the drag from the upper edges of the Earth's atmosphere and has not re-entered.

were being saturated by intense radiation from a radiation belt encircling the Earth, a finding later confirmed by satellites in the Explorer series. Pioneer spacecraft en route to the Moon, which they failed to reach in 1958, detected an outer belt of radiation. The doughnut-shaped radiation belts girdling earth were named after Van Allen.

Powered by its batteries, Explorer 1 continued transmitting until 23 May 1958, and remained in orbit until 1970. The very first satellite to carry an alternative to batteries was the first operational Vanguard spacecraft, launched into an orbit between 653km (406 miles) and 3897km (2421 miles) from earth on 17 March 1958. In addition to batteries, the 1.47kg (3.2lb) Vanguard 1, with a diameter of 16.25cm (6.4in), carried six small packs of silicon solar cells mounted around the sphere,

which were a precursor to the main source of electrical power for satellites of the future. Mirror-like solar cells converted the energy of the Sun into electricity, and in the case of Vanguard 1 were used to power the craft's five-milliwatt transmitter.

Today, Vanguard 1 is the oldest object still in space. With its relatively high period of orbit of two hours 13 minutes, the satellite has avoided the drag from the upper edges of the Earth's atmosphere and has not re-entered. It was succeeded by two other satellites of which Vanguard 3, launched in 1959, was the largest, a 50cm (1.64ft) diameter sphere weighing 23kg (51lb). The craft was designed to measure the Earth's magnetic field and radiation. It was placed in a 23° orbit of between 513km (320 miles) and 3524km (2190 miles). By the time it came to an end, the Vanguard

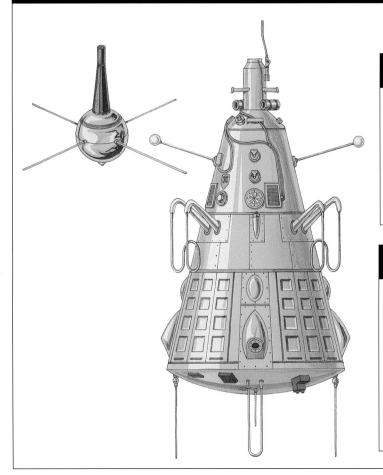

VANGUARD 3 AND SPUTNIK 3

Above: America's first satellite, Explorer 1, mounted on the very top of a Jupiter C booster, begins its historic orbital flight from Cape Canaveral. The Jupiter C was a Redstone IRBM with three solid-propellant upper stages.

Right: The American 22kg (50lb) Vanguard 3 satellite and the Soviet Sputnik 3. Vanguard 3 was launched in September 1959, 16 months after Sputnik 3 which weighed 1.3 tonnes.

VANGUARD 3

Diameter: 50.8cm (20in)
Weight, payload: 23kg (50.7lb)
Orbit: 512km x 3744km (318 miles x 2326 miles) x 33.35°

SPUTNIK 3

Length: 355cm (140in)
Base width: 173cm (68in)
Frequencies: 20.005 MHz; 40.008 MHz
Orbit: 217km x 1864km (135 miles x 1158 miles) x 65.18°

programme had suffered eight launch failures, a number unacceptable by today's standards.

Film of those early spectacular US launch failures are still regularly shown on TV, the most spectacular being the demise of a Beacon satellite aboard a Juno 2 rocket in 1959. The Juno 2 lifted off, turned left as it rose above the launch tower, turned left again towards the ground, and was blown apart by Cape Canaveral's range safety officer.

Surprisingly, after the startling beginning made by the Soviet Union, the only other major science satellite it launched during the first five years of the Space Age was Sputnik 3 on 15 May 1958. This satellite weighed 1.33 tonnes, operated for 692 days, and provided a wealth of data on a variety of geophysical and space physics phenomena. Many other Soviet launches, all designated as Sputnik, were dedicated to the development of a manned space programme. Others later emerged as cover-ups for failed Moon and planetary missions.

APPLICATIONS OF SPACE

The US, meanwhile, proceeded to launch dozens more satellites, increasing scientific knowledge and demonstrating important space applications. The first satellite dedicated specifically to observing the Sun, for example, was launched on 7 March 1962. Called the Orbiting Solar Observatory (OSO), it transmitted data on 75 solar flares over a period of 17 months. The 208kg (459lb) OSO was placed in an orbit of between 553km (344 miles) and 595km (370 miles), with an inclination of 32.9°.

Fourteen Explorer satellites had also been launched by the end of 1962, including the 64kg (141lb) Explorer 6, launched on 7 August 1959 into a 47° orbit of between 245km (152 miles) and 42,400km (26,346 miles), the highest apogee yet reached by a satellite. Explorer 6 was nicknamed the "paddle wheel satellite" because it carried an array of solar cells, not on the body of the spacecraft but on four panels attached to the satellite by vanes. The panels carried a total of 8000 solar cells capable of charging the craft's nickel-cadmium batteries. This allowed the satellite to carry more scientific instruments than ever, eight in total, one of which attempted to take the first scanned TV images of Earth. The resulting blurred images were a disappointment. Imaging of the Earth was the mission of the first major space applications programme, although few people realized it at the time.

On 28 February 1959, the US Air Force launched the 618kg (1362lb) Discoverer 1 into an orbit of 90° inclination, which took it over the poles of the Earth. It was powered by a Thor Agena booster from a new launch site at Vandenberg on the California coast. By orbiting in the range

The Discoverer 13 capsule which splashed down in the sea was the first object to be returned safely from orbit, an historic achievement. Further recoveries were made, principally by catching them in a net trailed by an aircraft.

SPACE RACE SCOREBOARD 1957-62

US: The US made 101 successful satellite and space probe launches, including two planetary missions; they experienced 30 launch failures including six of interplanetary missions. The best sequence of successful launches was 14; its worst sequence of failures was seven, including two interplanetary launches.
USSR: The Soviet Union made 31 successful launches, including three interplanetary flights, while experiencing 24 failures, including 13 on attempted interplanetary flights. Its best sequence of successful launches was nine, and its worst failure sequence was in four interplanetary launch attempts.

of 212km (132 miles) to 848km (527 miles), the satellite made about 17 orbits in one day while the Earth rotated beneath it, thereby flying over the whole of the Earth's surface in one day. A satellite in a lower orbital inclination of, for example, 30° would pass over areas on earth only to a limit of 30° north and south latitude.

Although Discoverer 1 was ostensibly on a research mission, it failed. After a number of subsequent failures, a capsule from Discoverer 13, equipped with a heat shield, was successfully recovered after a retrorocket was fired to slow its orbit down, to make a safe re-entry through the Earth's atmosphere. The Discoverer 13 capsule which splashed down in the sea was the first object to be returned safely from orbit, an historic achievement. Further recoveries of capsules were made, principally by catching them in a net trailed by an aircraft. The Discoverer programme continued up to number 38 in February 1962, when, for the first time in space history, the type of satellite used in a launch was made classified information by the US Air Force.

What would have been Discoverer 39 was kept under a security blanket and further launches continued in partial secrecy. It was not until the 1990s that the true nature of the Discoverer programme was confirmed. The Discoverers were actually Corona satellites carrying high-resolution cameras to spy on the Soviet Union. The capsules returned the valuable film to earth for processing. The spacecraft principally consisted of the Agena second stage equipped with a 70° panoramic Itek camera, with Eastman Kodak film, and loaded in the re-entry capsule. The original resolution (the amount of detail that could be seen) of the images taken by the satellites was about 10m (32.8ft), but by the time the programme ended in 1972, to make way for larger and more powerful satellites, the resolution had been reduced to 1m (3.2ft). Despite the public's assumption of an enormous Soviet military threat, the Corona satellites showed the US that the Soviet's bark was bigger than its potential bite, a fact not appreciated by the public at the time because the Cold War was intensifying.

Far left: A 21m (68.9ft) tall Vanguard rocket is launched from Cape Canaveral carrying a satellite of the same name. The Vanguard was based on a Viking sounding rocket, with a second stage based on an Aerobee. It was equipped with a solid-propellant upper stage.

Left: Known as the "paddle-wheel" satellite, Explorer 6 was the first to carry its electrical-power-generating solar cells on panels attached to its instrumented sphere on four booms. The satellite took the first crude images of the Earth from orbit.

A new Soviet programme called Cosmos was introduced in 1962, principally for scientific purposes but, like the Coronas, was later to be used primarily as cover for military projects, especially reconnaissance. However, two other military projects, called Samos and Midas, were openly revealed by the US. Prototype satellites for these projects were first launched successfully in 1960. Samos stood for Satellite and Missile Observation System and the satellite was based on the Corona Agena stage design. This time, however, instead of a film package, Samos carried a TV camera which was designed to transmit reconnaissance images directly to the ground, thereby saving the time it took to recover a Corona satellite capsule and develop its valuable film. Inevitably, the resolution of the TV camera could not match that of the optical

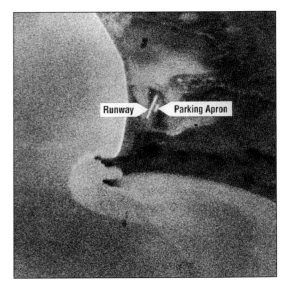

Corona cameras, but the Samos craft were prototypes of much more powerful systems of the future. The first successful Samos, No 2, which weighed 1.9 tonnes, was launched on an Atlas Agena booster

THE EXPLORER SATELLITES

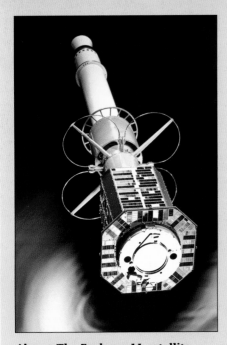

Above: The Explorer 11 satellite was launched in 1961 to detect high-energy gamma rays from cosmic sources, and map their distribution in the sky.

Date	Development
31 January 1958	Explorer 1 discovers Van Allen radiation belt.
26 March 1958	Explorer 3 returns radiation and micrometeorite data.
24 August 1958	Explorer 4 maps radiation belts.
7 August 1959	Explorer 6 flies for science and obtains first raw Earth images.
13 October 1959	Explorer 7 returns magnetic-field and solar-flare data.
3 November 1960	Explorer 8 studies the ionosphere.
16 February 1961	Explorer 9 monitors atmospheric density.
25 March 1961	Explorer 10 maps magnetic fields.
27 April 1961	Explorer 11 studies gamma rays.
16 August 1961	Explorer 12 returns solar wind and radiation data.
25 August 1961	Explorer 13 monitors micrometeorites.
2 October 1962	Explorer 14 maps the magnetosphere.
27 October 1962	Explorer 15 monitors radiation decay.
16 December 1962	Explorer 16 monitors micrometeorites.

Note: Explorers 2 and 5, and six other Explorer satellites fail.

PROJECT CORONA

After a series of failed launches and a number of demonstration flights, Discoverer 14's capsule was successfully caught in mid-air by a C-119 recovery aircraft with a special capture net. Inside it were the first exposed reconnaissance images taken from space. When the photo interpreters looked at the images they were jubilant. Imagery of over 2,640,000km (1,640,400 miles) of Soviet territory had been photographed, with a best resolution of about 10m (32.8ft). Eisenhower was astounded and ordered that the project be placed under a stricter security blanket, and that the capsule be destroyed to prevent it falling into the hands of spies! Earlier, when the test capsule from Discoverer 2 missed its landing point and came down in the Arctic, a massive search was launched for it amid fears that it would fall into enemy hands. The vain search for the capsule formed the basis of the Alastair Maclean novel and film *Ice Station Zebra*. It was obvious from the Discoverer images that the Soviet missile threat was not nearly so serious as feared. However, the US Government perpetuated the threat, promoting an arms build-up and an accelerated military missile programme. By the end of the Corona programme, satellites had imaged all Soviet missile bases, imaged each Soviet submarine class, revealed the presence of Soviet missiles protecting the Suez Canal in Egypt, and identified Soviet nuclear assistance to the People's Republic of China.

Far left: The very first image taken by a spy satellite. This view of Mys-Shmidta Air Field in the USSR was taken on 18 August 1960 by Discoverer 14.

Below: The instrumented, recoverable re-entry capsule of a Discoverer Corona Project satellite is prepared for launch. It is equipped with a 70° panoramic Itek camera with Eastman Kodak film.

on 31 January 1961, and entered a 97° inclined orbit between 474km (295 miles) and 557km (346 miles) from Earth.

Midas was also based on the Agena spacecraft "bus". Its name stood for Missile Defense Alarm System. The Midas satellite payload was an infrared sensor which was designed to detect the heat radiated by the exhaust of a missile or rocket launch. The theory at the time was that an eventual early warning system could detect the launch of a missile from the Soviet Union heading for the US. The first 2.3 tonne Midas to reach orbit was also launched by an Atlas Agena, on 24 May 1960. Midas 2 reached a 33° inclination, with an orbit of between 484km (301 miles) and 511km (319 miles). However, it failed the following day. It was more of a prototype than an operational vehicle. Midas 3 followed on 12 July 1961, launched from Vandenberg rather than Canaveral. It weighed 1.6 tonnes, and entered a 91.1° inclination, with an almost perfectly circular orbit of between 3345km (2078 miles) and 3538km (2198 miles).

THE DISCOVERER PROGRAMME

DISCOVERER CAPSULE

The re-entry capsule, built by General Electric Co. weighed about 136kg (300lb). The part recovered by parachute was enclosed within a protective heat shield. Capsule contents varied from launch to launch; Discoverer 14, launched 18 August 1960, was the first to carry film from a reconnaissance camera. The capsule of Discoverer 32 investigated the effect of radiation on (a) metal samples; (b) genetic properties of seed corn; (c) shielding materials, and (d) silicon solar cells. Discoverer 36 had a biopack with human and animal tissues, spores and algae.

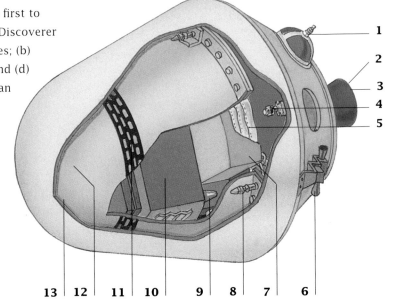

1 Cold gas storage tank
2 Thrust cone
3 Retro-rocket
4 Explosive belt
5 Recovery parachute and chaff
6 Stabilization jets
7 Parachute cover
8 Explosive pistons
9 Flashing light
10 Instrumentation package
11 Dye markers
12 Recovery capsule
13 Radio beacon (inside)
14 Ablating re-entry shield

DISCOVERER RECOVERY

The Discoverer spacecraft released its capsule at a backward angle while passing over Alaska. After re-entering the atmosphere protected by a disposable heat shield, the capsule was recovered by parachute near Hawaii. A patrolling C-119 (later C-130) aircraft snagged the parachute as it descended. If the capsule was missed, it floated just below the surface of the sea.

1 Pitch down
2 Separation and retro-fire
3 Re-entry heating
4 Recovery by patrolling C-119 or C-130 aircraft

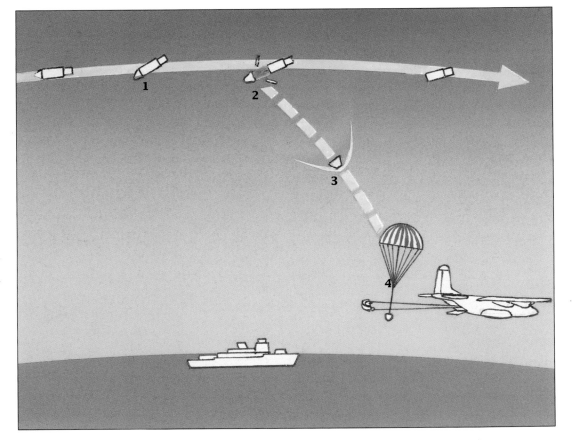

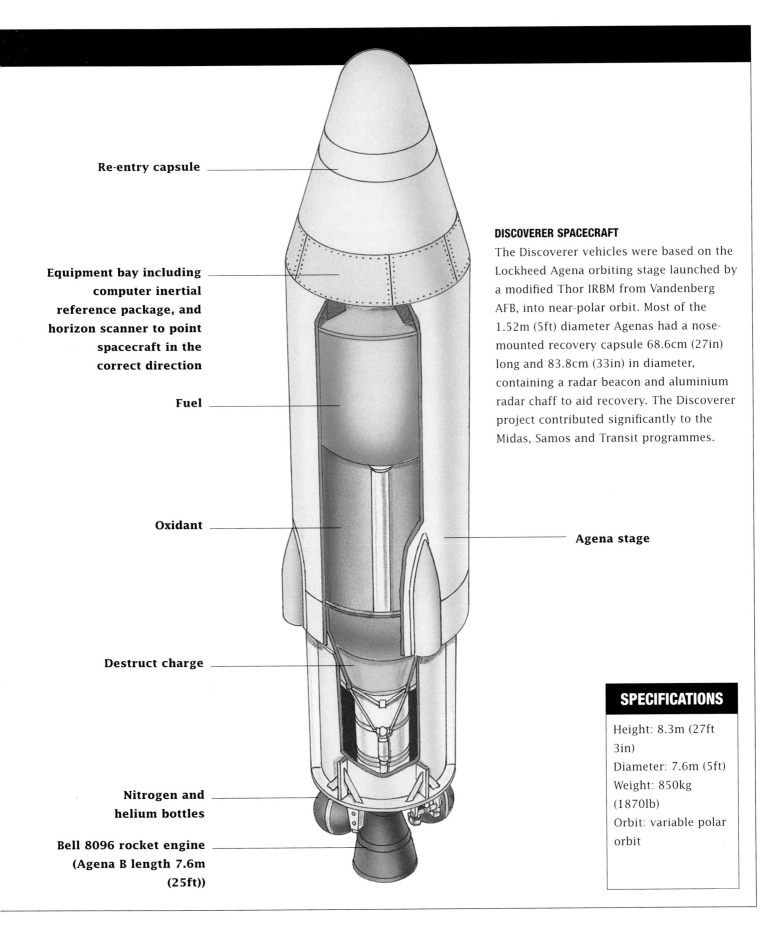

Re-entry capsule

Equipment bay including
computer inertial
reference package, and
horizon scanner to point
spacecraft in the
correct direction

Fuel

Oxidant

Destruct charge

Nitrogen and
helium bottles

Bell 8096 rocket engine
(Agena B length 7.6m
(25ft))

DISCOVERER SPACECRAFT

The Discoverer vehicles were based on the
Lockheed Agena orbiting stage launched by
a modified Thor IRBM from Vandenberg
AFB, into near-polar orbit. Most of the
1.52m (5ft) diameter Agenas had a nose-
mounted recovery capsule 68.6cm (27in)
long and 83.8cm (33in) in diameter,
containing a radar beacon and aluminium
radar chaff to aid recovery. The Discoverer
project contributed significantly to the
Midas, Samos and Transit programmes.

Agena stage

SPECIFICATIONS

Height: 8.3m (27ft
3in)
Diameter: 7.6m (5ft)
Weight: 850kg
(1870lb)
Orbit: variable polar
orbit

Above: An Atlas Agena booster launches a Samos satellite into orbit from Point Arguello, California.

operational system yet. This was confirmed later when it was revealed that the first fully successful Midas mission was not flown until 1963.

The next space application to be introduced was navigation. The US Navy launched the Transit 1B in April 1960 to demonstrate the ability of satellites to provide navigation services. The 121kg (267lb) spacecraft, launched on a Thor Able Star booster from Cape Canaveral, was placed into an orbit of between 373km (232 miles) and 478km (297 miles), and inclined to the equator at 51°. The Transit 1B was the precursor of a planned three-satellite system designed to provide the first space-based navigation for ships and submarines of the US Navy. The best predicted positioning accuracy was to within 160m (524ft). During the first five years of the Space Age, only prototype Transit navigation satellites were launched, but they provided much of the experience needed to establish an operational system.

Perhaps the space application that proved its worth much earlier than others was weather forecasting, utilizing the Tiros satellite system. Short for Television and Infra Red Observation Satellite, the first Tiros was launched on 1 April 1960 into an orbit of between 677km (421 miles) and 722km (449 miles) and inclined at 48°. By the end of 1962, six satellites had been launched. Shaped rather like a hat box, 106cm (3.4ft) in diameter and weighing 129kg (284lb), Tiros was covered with 9260 solar cells which recharged 64 batteries. It was equipped with wide-angle and narrow-angle high-resolution vidicon cameras, each with magnetic tape recorders capable of both storing 32 images during each orbit, and transmitting the images when the craft came to within 2500km (1553 miles) of a ground station.

The plastic tape of the recorders was 42m (138ft) long, and moved at a speed of 127cm (4.1ft) per minute during playback. The Tiros satellite was oriented carefully in

Midas 4 proved the technology by detecting the launch of a Titan 1 missile from Cape Canaveral in October 1961, although the detection was reported some 90 seconds after launch. Clearly Midas was not an

relation to the horizon by using an infrared sensor, and was stabilized by spinning at a rate of 9rpm. The first three Tiros satellites to be launched returned 70,650 images alone. Although the images seem rather hazy when compared with today's weather pictures, the Tiros satellites were highly successful precursors to modern weather satellites, and performed effectively, particularly in giving early warning of the path of hurricanes.

THE BEGINNING OF THE COMMUNICATIONS REVOLUTION

Communications was another technical application that was demonstrated by satellites during the first five years, including the legendary Telstar in 1962. The first demonstration of the use of a satellite to transmit messages, rather than radio signals, was made in December 1958, when an entire Atlas B booster, weighing 3.96 tonnes, was placed at an inclination of 32.3°, in an orbit of between 184km (114 miles) and 1483km (921 miles).

The satellite carried a payload called the Signal Communications Orbit Relay

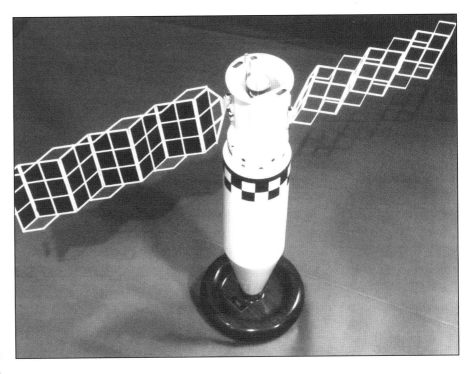

Experiment (SCORE), which was designed to pick up radio messages from ground stations, and deliver them to other ground stations on demand. SCORE relayed a Christmas greeting message from President Eisenhower during its 34 days in orbit.

In 1959, using a ready-made satellite –

Above: The Missile Defense Alarm System (MIDAS) satellite was equipped with unfurlable solar panels to provide the spacecraft with electrical power.

Left: The Transit 1B was the prototype of the first navigational satellites.

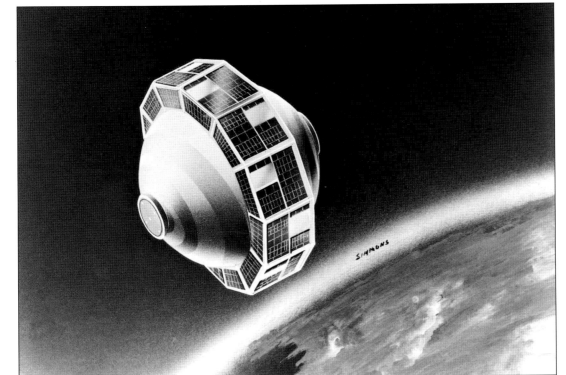

THE TELSTAR REVOLUTION

European stations received some test TV transmissions before the first trans-Atlantic TV programme was broadcast via satellite by France to the US at 19:35 hours, US time, on 11 July 1962. The French Communications Minister, Jaques M. Marette, acted as the master of ceremonies on the programme recorded earlier and said, "Relax. You are in Paris and I invite you to spend a few pleasant moments with me". He introduced a French film star, Yves Montand, who sang "La Chansonette" (The little song). The seven-minute programme caused ill-feeling as it was meant to have been a joint effort with Britain. France claimed it was only a test, not a programme. However, a few hours later

Above: President John F. Kennedy featured in the first intercontinental TV programme broadcast live by the Telstar satellite.

Britain claimed the distinction of sending the first live pictures from Europe to the US, showing the Goonhilly station. On 23 July came the most spectacular communications event in history at that point. Sixteen European countries exchanged TV programmes with the US, as more than 200 million viewers in Europe and the US saw daytime US and night-time Europe. The audience was taken through Europe, from Lapland to Sicily, and "travelled" across America, from the World's Fair in Seattle to the Statue of Liberty in New York Harbour. President Kennedy addressed the TV audience live from the White House, astronaut John Glenn talked about the Mercury space programme, and astronaut Wally Schirra talked about his forthcoming mission in Sigma 7.

Below: The 130kg (287lb) Tiros satellite was equipped with 9260 solar cells providing power to recharge the craft's 64 batteries. The satellite carried wide-angle and narrow angle videcon cameras which took 32 images during each orbit.

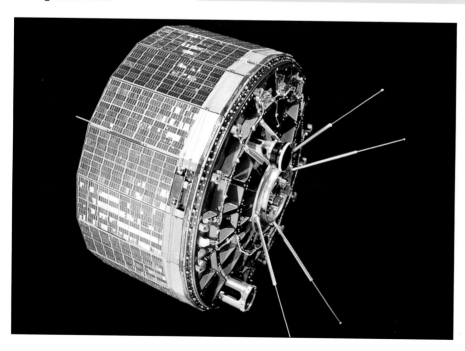

the Moon – radio signals were transmitted from Jodrell Bank in the UK and bounced off the lunar surface to be received three seconds later in Cambridge, Massachusetts. This method of communications experiment was further utilized with an orbiting 30m (98ft) diameter balloon, called Echo 1, which reflected voice and TV signals from Holmdel, New Jersey, to Goldstone, California.

Echo 1 failed to orbit but Echo 1B was launched successfully on 12 August 1960 into a 75.9° inclination, with an orbit of between 1524km (947 miles) and 1684km (1046 miles). It consisted of a Mylar plastic skin covered with vapour-deposited aluminium which had a thickness of just

Left: The Signals
Communications Orbit Relay
Experiment (SCORE)
instrument was carried into
orbit aboard an entire Atlas B
rocket in December 1958, and
relayed a recorded Christmas
greeting message from the US
President.

0.00127cm (0.0005in). When it was deployed from the second stage of its Thor Delta booster, the 61kg (134lb) balloon was inflated by means of around 20kg (44lb) of acetamide, a sublimating powder placed within the satellite before it was packed. As the balloon unfolded in space, the powder turned to gas and expanded the structure to its full diameter of 30m (98ft) in about 10 minutes. Because it was highly reflective, the Echo balloon could be seen clearly in the sky overhead like a slow moving star, and captured the imagination of the world. The times for sighting the satellite were regularly posted in newspapers.

The most significant advance in space applications occurred when the 77kg (170lb) satellite, Telstar 1, was launched on 10 July 1962. Sent into an orbit inclined at 44.8°, of between 936km (582 miles) and 5653km (3512 miles), it transmitted the first live TV pictures between the US and Europe, beginning the satellite communications revolution. Its orbit enabled it to be "seen" by ground stations in both the US and Europe at the same time as it passed over the Atlantic Ocean.

Telstar 1 was a 0.8m (2.6ft) diameter sphere which was operated by the American Telephone and Telegraph Company to demonstrate broadband TV and telephone communications via satellite. Telstar was the precursor to the geostationary orbiting communications satellites of today. The key technology on Telstar was a travelling wave-tube amplifier and associated equipment, which enabled a signal transmitted from the Bell Company's ground station in Andover, Maine, to be amplified by 5000 times, so that it could be received in ground stations at Goonhilly Downs,

Right: The 30m (98ft) diameter Echo 1 balloon had a Mylar plastic skin covered with highly reflective vapour-deposited aluminium. The balloon was carried in a tightly packed canister and inflated in space.

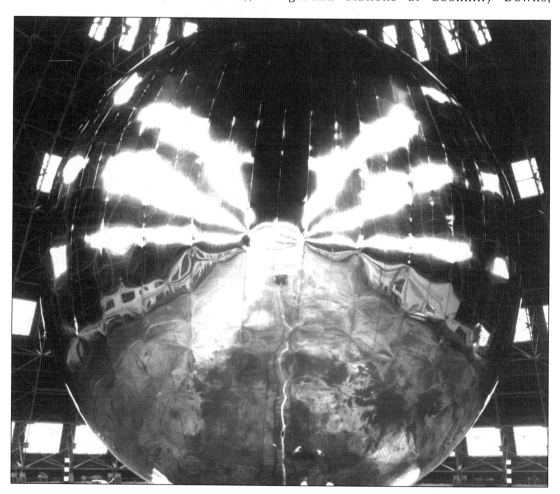

Left: The pioneering Telstar satellite, launched in July 1962, beamed the first live TV across the Atlantic Ocean between the US and Europe, beginning a communications revolution that has led to the satellite TV of today.

Below: The UK's first satellite, Ariel 1, was launched in April 1962, aboard a US Thor Delta booster. It carried an array of scientific instruments designed to study the Earth's ionosphere and cosmic rays.

Cornwall in the UK and Pleumer-Bodou, Brittany in France. The first live picture was transmitted between the US and Europe on 11 July, and it was eventually followed by live TV programmes, including a programme linking locations and events on the two continents, as well as a live message from President Kennedy at the White House.

The first non-US, non-Soviet country to have a satellite in space was the UK with the 60kg (132lb) Ariel 1. Ariel 1 entered an orbit at 53.9°, between of 389km (242 miles) and 1214km (754 miles) from Earth, on 26 April 1962. Ariel 1 was designed to investigate the Earth's ionosphere and its relationships with the Sun, and to record heavy primary cosmic rays.

By December 1962, extraordinary advances had been made in space technology. There was an increased understanding about Earth's place in space and its interaction with its space environment. The beginnings of the exploitation of space, which had not been considered possible when Sputnik 1 started the Space Age five years before, had begun in earnest.

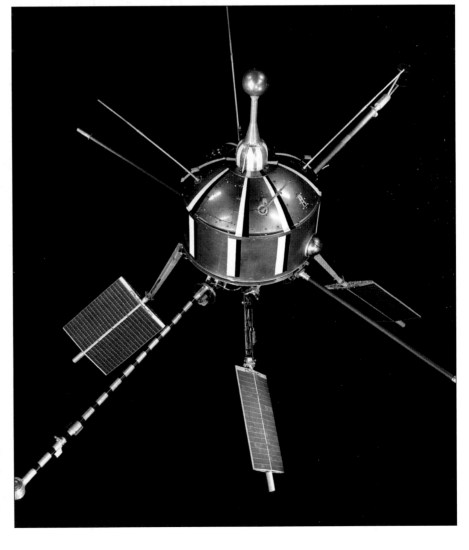

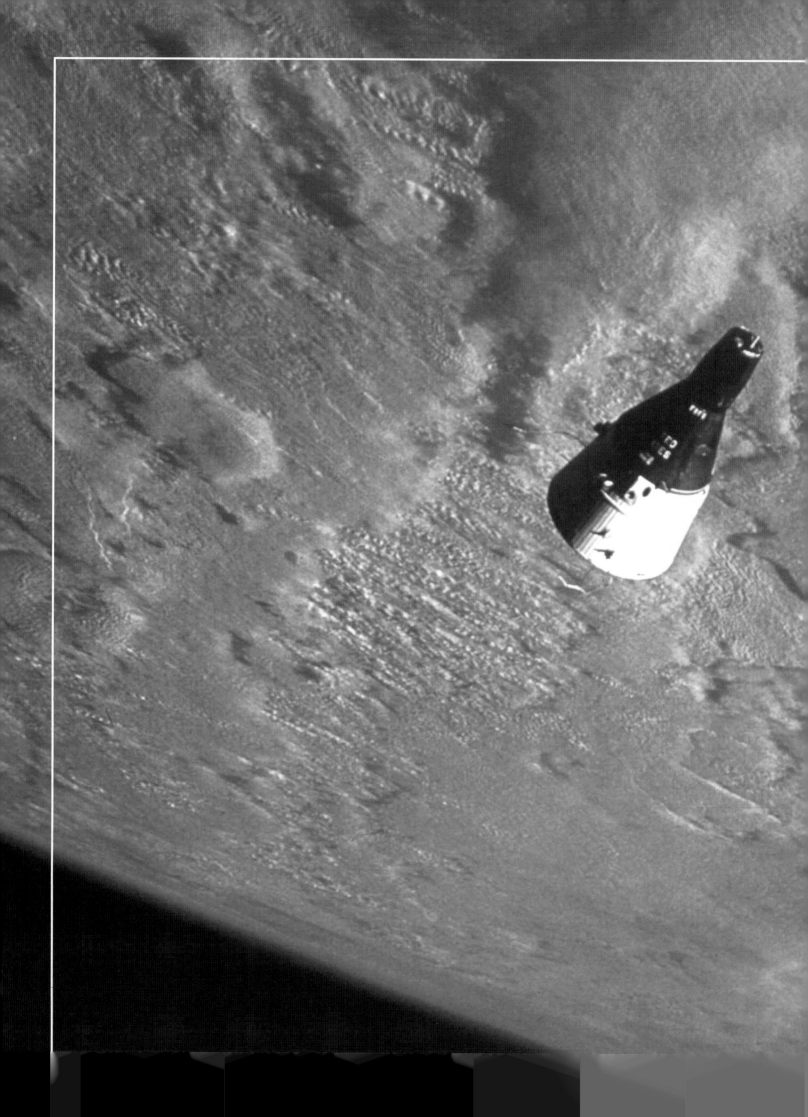

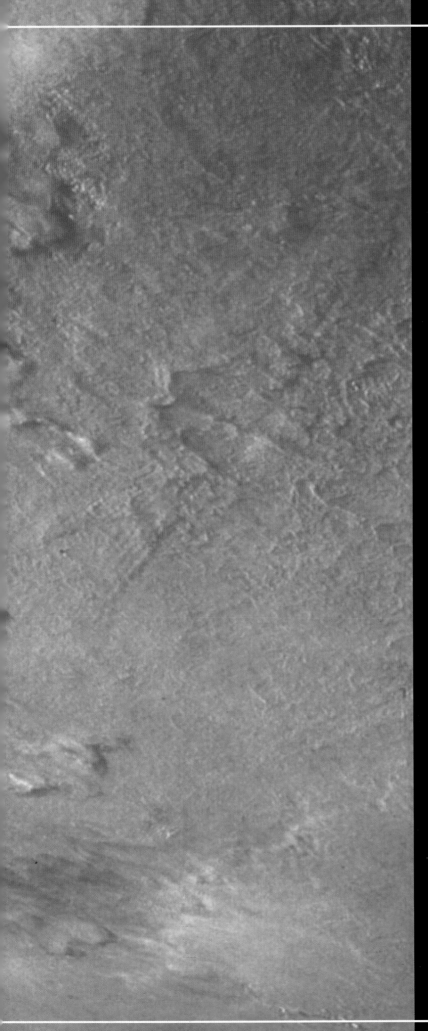

THE FIRST MANNED SPACE VEHICLES

Once the Space Age had begun it was inevitable that one day men would be launched into space. In fact it took only four years before the first space traveller was in orbit. This extraordinary rush into space was the result of the Cold War rivalry between the superpowers, in which it became a race to get the first man into space.

The only way to launch people into space so quickly was to use existing rockets, such as the US and Soviet ICBMs. And the best way in which to fly, given the urgency, was to place a spaceman in the nose cone of a rocket. The payload had to be more sophisticated, of course, to sustain a human life on board and bring the traveller back to earth in one piece, but nonetheless the spacecraft were very basic and a far cry from the futuristic space planes that

Left: Gemini 7 is photographed from Gemini 6 during the historic first space rendezvous which took place in December 1965.

VOSTOK

The Vostok spacecraft on its Vostok booster. The craft consisted of a conical instrument unit, including the retro-rocket system, and the spherical crew module in which the cosmonaut lay in a contour couch, mounted on an ejector seat.

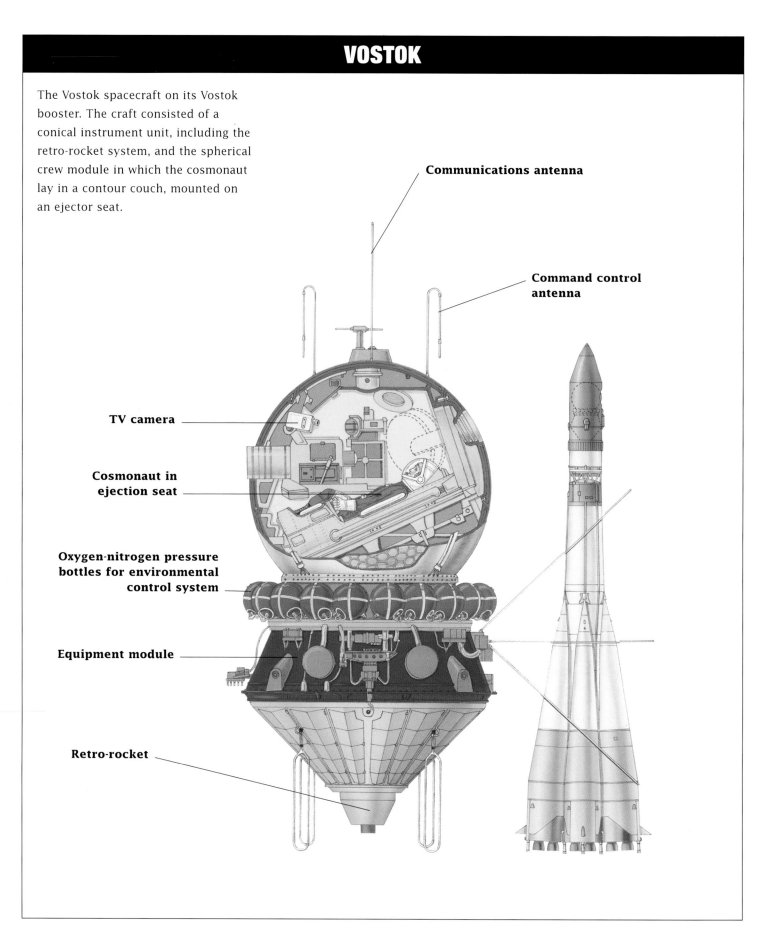

Communications antenna

Command control antenna

TV camera

Cosmonaut in ejection seat

Oxygen-nitrogen pressure bottles for environmental control system

Equipment module

Retro-rocket

Left: The Vostok booster leaves the Baikonur launch pad No. 1, carrying Yuri Gagarin towards Earth orbit. The first pictures and spectacular film of the launch were not revealed until 1968.

were being imagined, even before Sputnik 1 was launched.

With the Soviet Union having launched the first satellite and appearing to lead the space race hands down, the prize of the first manned space flight was eagerly sought in 1959. Both superpowers began their quests by designing and testing the manned spacecraft, and selecting a team of military pilots as the first space travellers. They became known as cosmonauts in the Soviet Union and astronauts in the US. Both nations conducted several test flights of models of their spacecraft, which became known as Vostok for the Soviets, and Mercury for the US. The US programme was entirely open, the Soviet Union's top secret. These various test flights prepared the way for the launchers and spacecraft, and a number of rockets carried animals as pathfinders. While the Soviets used dogs, the US astronauts were particularly irked by the fact that they were being preceded in space by chimpanzees. But these tests were invaluable and vital, especially since

many ended in failure – in some Soviet cases, in the deaths of the animals – although this was not revealed at the time.

VOSTOK

By 1961, both nations were ready to launch their first space traveller. US Navy Commander Alan Shepard was preparing for a sub-orbital, up-and-down test flight into space in March 1961, but this was delayed to May when it was decided to fly one more unmanned Redstone-boosted test flight. On 12 April 1961, the Soviet Union launched Air Force Captain Yuri Gagarin into orbit. Shepard and the US had been beaten. Gagarin's flight was one of the greatest journeys of the 20th century, receiving extraordinary press coverage worldwide. Gagarin was feted as a world hero. The Soviet Union had won the latest race in space. In reality, Gagarin had been a passenger and did not even land back on Earth in his spacecraft. Vostok was launched on the SL-3, a booster similar to the ICBM that launched Sputnik, but with an additional upper stage from which the

A manned space flight was strictly defined as one in which a passenger flew in a craft as it was launched and as it landed. The fact that Gagarin had landed by parachute separately from his craft was not revealed until long after his historic flight.

Vostok separated after reaching its predesignated orbit.

The 4.73 tonne Vostok spacecraft was 4.4m (14.4ft) long and 2.43m (8ft) in diameter. The cosmonaut flew in a spherical descent module which weighed 2.46 tonnes and had a diameter of 2.3m (7.5ft). The seat of the spacecraft was an ejection seat to allow the cosmonaut to exit the craft which, even under a parachute, impacted at a speed of about 10m (32.8ft) per second, with a fatal 100g load. Coming down separately under parachute, the cosmonaut landed at about 5m (16.4ft) per second. A much larger parachute system or even soft-landing retrorockets were not possible owing to launch weight constraints. The ejection seat would also be used in the event of a launch failure. Even then, the flight was very simple and the cosmonaut was basically a passenger. The cabin was fitted with a food locker, radio, experiment cabinet, and a porthole with an optical orientation indicator. There was enough room in the cockpit for the cosmonaut to undo straps and float around. Gagarin did not do so, but later cosmonauts did.

The module also had external command, control and communications antennae. The all-round outer heat shield of the capsule was designed to absorb the heat, and burn away during re-entry. Beneath the descent module was an instrument module, attached to it by umbilicals and four metal straps which extended round the descent module. The 2.27 tonne instrument section was 2.25m (7.4ft) long and 2.43m (8ft) at its maximum diameter. It provided the oxygen and nitrogen for the cosmonaut's life support system and also, at its conical base, the vital

GAGARIN'S ESCAPE

Apart from his many achievements, Yuri Gagarin could have been the first man to die in space. It was revealed much later that Gagarin's spacecraft malfunctioned after retro-fire and could have been destroyed during re-entry. Inside his spherical capsule, Gagarin could feel the "kick" as the retrorocket on the attached service module fired to slow the craft down for re-entry. The retrorocket fired for 40 seconds as planned. The instrument section was designed to be jettisoned, leaving the circular capsule to plunge into the atmosphere, but it failed to separate. There was a sharp jolt and the spacecraft went into a 30rpm spin in all three axes. As the craft entered the upper layers of the Earth's atmosphere, the rotation abated and became more of a 90° right-to-left oscillation. Gagarin noticed a bright crimson light appearing out of his window. Re-entry had begun, with loud crackling noises. The spacecraft was making an uncontrollable re-entry and there was an acute danger of disintegration. G loads built up to a level of 10g. It was only when the heat became so intense that it severed an umbilical between the capsule and the instrument section that the latter fell away. Gagarin's re-entry then stabilized. He was ejected automatically as planned, and he landed in a

Above: Yuri Gagarin on the launch pad at the Baikonur Cosmodrome on 12 April 1961 bids farewell to Sergei Korolev.

ploughed field, watched by a bemused woman and child who had to be assured that he was not an "alien". According to Federation Aeronautique International rules, a manned space flight was to be one in which a passenger flew in a craft as it was launched and as it landed, which is what the Soviets claimed. The fact that Gagarin had landed by parachute separately from his craft was not revealed until much later.

SOVIET VOSTOK MISSIONS

Above: The dogs Strelka and Belka are held aloft in triumph after becoming the first living creatures to return alive to Earth from orbit in August 1960.

Above: Valentina Tereshkova became the first woman in space, orbiting aboard the Vostok 6 in June 1963.

Date	Vehicle	Mission
18 July 1959	Vostok	Unmanned test, launch fails.
15 April 1960	Vostok	Unmanned test, launch fails.
16 April 1960	Vostok	Unmanned test, launch fails.
15 May 1960	Sputnik 4	Unmanned test with dummy cosmonaut. Retrorocket sends satellite into high orbit instead of re-entry.
28 July 1960	Vostok	Dogs Chaika and Lisichka killed as rocket explodes.
19 August 1960	Sputnik 5	Dogs Strelka and Belka recovered from orbit in Vostok capsule after 18 orbits.
1 December 1960	Sputnik 6	Vostok is incinerated during re-entry. Dogs Pchelka and Mushka killed.
22 December 1960	Vostok	Dogs Damka and Krasavk, recovered after launch failure.
9 March 1961	Sputnik 9	Dog Chernushka recovered after Vostok test.
25 March 1961	Sputnik 10	Dog Zvezdochka recovered after Vostok test.
12 April 1961	Vostok 1	Yuri Gagarin first cosmonaut, 1hr 48min.
6 August 1961	Vostok 2	Gherman Titov makes first day-long flight lasting 1 day 1hr 11min.
11 August 1962	Vostok 3	Andrian Nikolyev makes flight lasting 3 days 22hr 9min.
12 August 1962	Vostok 4	Pavel Popovich passes close to Vostok during flight lasting 2 days 22hr 44min.
14 June 1963	Vostok 5	Valeri Bykovsky flies longest solo mission of 4 days 22hr 56min.
16 June 1963	Vostok 6	Valentina Tereshkova is the first woman in space. Flight lasts 2 days 22hr 40min.

retrorocket to slow down the craft for re-entry into the Earth's atmosphere. The Vostoks were launched into orbits low enough to ensure that gravity and atmospheric drag would cause a natural re-entry within ten days if the retrorocket happened to fail. The 1.61 tonne thrust rocket used hypergolic nitrogen tetroxide oxidizer and an amine-based fuel which ignited spontaneously, thus not requiring an ignition source. It fired for 45 seconds, reducing the orbital velocity by about 155m (509ft) per second.

THE MERCURY PROGRAMME

On 5 May 1961, soon after Gagarin's landing, the eager Shepard roared into space for the US aboard a Redstone rocket carrying his Mercury capsule, called Freedom 7. He went on a sub-orbital flight lasting only 15 minutes in which he proved that an astronaut could take

with an ablative heat shield, and the capsule had to be carefully orientated during re-entry so that the heat shield was pointing in the right direction, and making the right angle of entry into the atmosphere. The solid-propellant retrorocket was attached to the heat shield, and was normally deployed after firing and before re-entry. As Mercury was a piloted vehicle, unlike Vostok, its cockpit contained more than 100 displays and consoles which were mounted in front of the astronaut's face, and provided spacecraft orientation, navigation, and environmental and communications control. The central section contained a periscope, and the capsule had a rectangular window on all but the first manned flight, which had a circular porthole.

The position of the Mercury capsule could be altered and controlled by the use of an airplane-like joystick commanding the release of short bursts of hydrogen peroxide gas from 10 control thrusters located on various parts of the craft. These movements could be controlled by the automatic stabilisation and control system (ASCS), the rate stabilisation and control system (RSCS), or by an operation, part manual and part electrical, known as fly-by-wire. The Mercury launch escape system comprised a cluster of solid propellant rockets on a tower positioned on top of the capsule. The entry-exit hatch could be blown open explosively in the event of an urgent evacuation, such as after splashdown. Mercury descended into the sea under a drogue parachute and a main parachute, and then deployed a landing bag from its rear to cushion the splashdown, which would otherwise have caused a 10g jolt.

Manned Mercury missions flew on the Redstone IRBM for suborbital flights, and the Atlas ICBM for orbital flights. The original plan was that each of the seven Mercury astronauts would fly a Redstone

Above: Mercury Redstone 3 (MR3) lifts off from Pad 5 at Cape Canaveral on 5 May 1961, carrying Alan Shepard into space aboard Freedom 7. His 15 minute flight was America's first manned space venture.

control of the attitude of the spacecraft in space and splashdown at sea, rather than on land as the Russian Vostoks had done. Shepard was feted as a hero by the US, but it was a fairly modest affair compared with Vostok 1. The Mercury capsule was also much smaller than Vostok, so small that astronauts joked that they did not get into it, they put it on.

The capsule was 2.76m (9ft) high with a maximum diameter across the base of 1.85m (6ft 1in). It weighed about 1.35 tonnes at launch. The base was covered

MERCURY

The confined space inside the Mercury capsule is clear in this illustration, showing the launch-escape-system rocket at the top of the spacecraft.

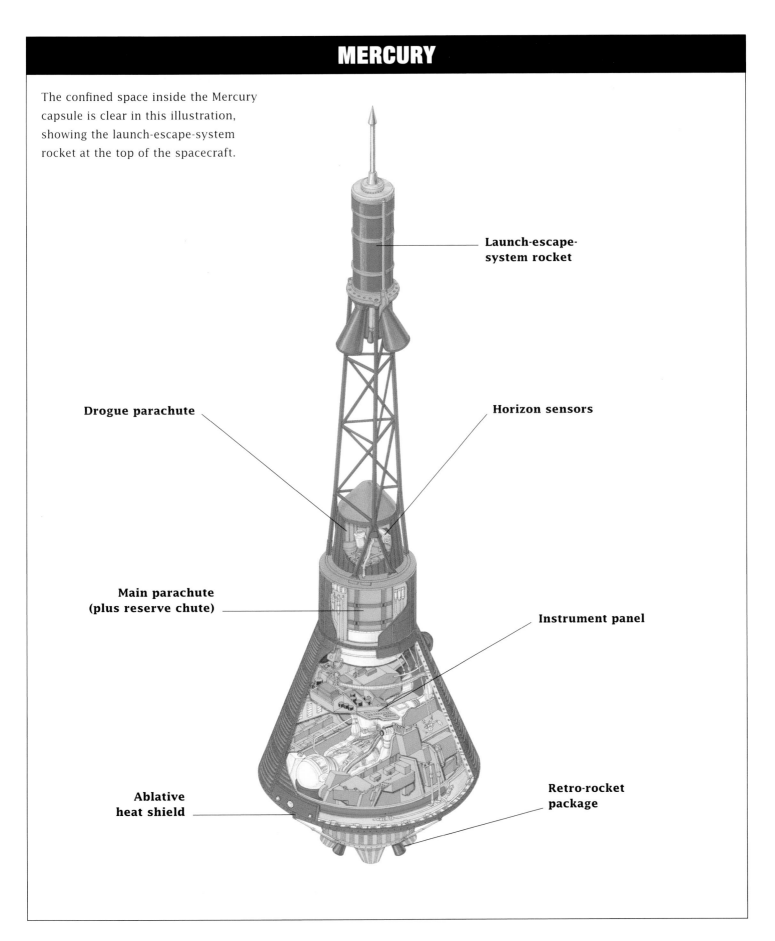

Launch-escape-system rocket

Horizon sensors

Drogue parachute

Main parachute (plus reserve chute)

Instrument panel

Ablative heat shield

Retro-rocket package

US MERCURY MISSIONS

Date	Vehicle	Mission
21 August 1959	Little Joe 1	Max Q abort-escape system test: failure.
9 September 1959	Big Joe	Ablation heat shield test: failure.
4 October 1959	Little Joe 6	Capsule aerodynamics test: partial success.
4 November 1959	Little Joe 1A	Max Q abort-escape system test: failure.
4 December 1959	Little Joe 2	Monkey Sam in high-altitude escape test: success.
21 January 1960	Little Joe 1B	Monkey Miss Sam in Max Q abort-escape test: success.
9 May 1960	Beach Abort	Launch-escape-system test: success.
29 July 1960	MA 1	First booster/spacecraft integrated launch: failure.
8 November 1960	Little Joe 5	Qualify capsule at Max Q: failure.
21 November 1960	MR 1	Aborted lift-off and accidental escape tower firing: failure.
19 December 1960	MR 1A	Suborbital flight recovery in 15min 45sec: success.
31 January 1961	MR 2	Flight of chimpanzee Ham: success but incident-filled.
18 March 1961	Little Joe 5A	Max Q escape and impact test: partial success.
24 March 1961	MR BD	Sub-orbital test: success.
25 April 1961	MA 3	Orbital test of Mercury capsule: booster exploded
28 April 1961	Little Joe 3B	Max Q escape sequence test: partial success.
5 May 1961	MR 3	Alan Shepard rides Freedom 7 on sub-orbital flight of 15min 28sec: success.
21 July 1961	MR 4	Gus Grissom rides Libertybell 7 on sub-orbital flight of 15min 37sec: success, but capsule sinks and Grissom is saved.
13 September 1961	MA 4	First Mercury capsule on one-orbit test flight: success
1 November 1961	MS 1	Mercury test: failure.
29 November 1962	MA 5	Flight by chimpanzee Enos, in Mercury capsule aborts after two orbits. Enos is recovered.
20 Feb 1962	MA 6	Three-orbit mission by John Glenn aboard Friendship 7 in 4hr 55min: success.
24 May 1962	MA 7	Three-orbit flight by Scott Carpenter aboard Aurora 7 in 4hr 56min: success.
3 October 1962	MA 8	Six-orbit flight by Wally Schirra aboard Sigma 7 in 9hr 13min: success.
15 May 1963	MA 9	22 orbit flight by Gordon Cooper aboard Faith 7 in 34hr 19min: success.

Above: The monkey Sam is prepared for his high-altitude test of the Mercury escape rocket.

Above: Gordon Cooper's Mercury Atlas vehicle sits on Pad 14 at Cape Canaveral ready for launch.

sub-orbital familiarisation flight before an orbital flight was attempted, but the plan was revised to just three sub-orbital flights. The second of these flights was made by Gus Grissom on 21 July 1961, aboard a capsule which he named Libertybell 7. The flight was another modest success until after splashdown, when the hatch was prematurely blown off. As water rushed in, Grissom scrambled out fast, almost drowning in the ensuing confusion as helicopters tried to prevent the capsule sinking. Thankfully Grissom was saved but the capsule sank. In 1999, Libertybell 7 was raised from the floor of the Atlantic Ocean after being spotted by a submersible search craft. The plan to fly further Redstone sub-orbital missions was scrapped.

MORE SOVIET 'FIRSTS'

Next to go was 25-year-old Soviet cosmonaut Gherman Titov, who is today still the youngest space traveller. He was launched into space aboard Vostok 2 on 6 August 1961, and remained there for more than a day, an extraordinary leap in terms of experience, given that Gagarin had made only one orbit. The main reason for Titov remaining in space for so long was that he had to land in the prime recovery zone – in the Soviet Union – in daytime, which would only be reached after 17 orbits because of the Earth's rotation. It was revealed later that Titov was spacesick but did manage to sleep for a while, and to eat and drink through toothpaste-like tubes. The spacesickness was caused mainly by the effect of weightlessness on the inner ear. Needless to say, the difficulties were not referred to

Left: The recovery of the Friendship 7 capsule which splashed down into the ocean after a nail-biting re-entry.

THE FRIENDSHIP EFFECT

John Glenn's mission aboard Friendship 7 on 20 February 1962 became one of the most famous in history, not so much because he made three orbits – a Russian cosmonaut had made 17 orbits five months previously – but because in Glenn, America had found a space hero. It was tough on the sub-orbital astronauts Alan Shepard and Gus Grissom who had preceded him in space. The fact that Glenn made the first US orbits was an enormous morale booster to the American public who quickly took to the smiling "boy next door" personality of the crew-cut, ginger-haired, freckle-faced astronaut. His flight received mammoth coverage in the press, and on live TV and radio around the world. To add drama to the mission, a signal, proved later to be erroneous, had indicated that Glenn's heat shield may have come loose, and that he and his Friendship 7 capsule might have been incinerated during re-entry. It was decided that he should keep his retrorocket pack on during re-entry in order to help keep the heat shield in place. But the entire event turned into a nail-biter for the public, and quite a spectacle for Glenn, as straps from the heat shield broke away and flew past him in flaming chunks. After the predicted communications blackout during re-entry, Glenn finally came back on the air amid the transmission crackles..."Boy, that was a real fireball!"

Above: An automatic camera aboard Friendship 7 monitors the astronaut John Glenn during the space flight.

Above: Wally Schirra's launch aboard Sigma 7 on 3 October 1962. His Atlas booster developed an alarming roll during the launch and an abort was considered.

(250 miles) – gave NASA the confidence to increase the length of missions.

Before the next Mercury flight could be attempted, however, the Soviet Union performed what on paper looked like the most spectacular feat in space history – two Soviet cosmonauts "meeting" in space. Vostok's Andriyan Nikolayev was launched first on 11 August 1962, and a day later Vostok 4 was launched crewed by Pavel Popovich. The two craft passed to within 6.4km (4 miles) of one another as their orbits coincided. This was not actually a meeting, neither was it a true rendezvous by two manoeuvrable craft changing their orbits. The Soviets and the gullible western press lapped it up and so the Soviet's "lead" in the Space Race was apparently extended further. Nikolayev, meanwhile, had extended manned space-flight time to almost four days.

Astronaut Wally Schirra extended the Mercury orbital experience to nine orbits aboard Sigma 7 on 3 October 1962, and was followed on 15 May 1963 by the last Mercury mission. This 34-hour flight by astronaut Gordon Cooper, who made almost 22 orbits, was a real triumph for the US, particularly as Cooper cheerfully overcame several problems in order to return home safely. The Soviet curtain-call for the Vostok programme was made in June 1963 and, not surprisingly, featured yet another first. Valentina Tereshkova became the first woman to travel in space on 16 June when she was launched on Vostok 6. She made a brief fly-by close to Vostok 5, which was manned by Valeri Bykovsky who had been launched two days earlier. Tereshkova, a former cotton-mill worker and amateur parachutist had an unhappy experience, however, feeling sick for most of her journey through space and even pleading to come down earlier than planned. Bykovsky on the other hand made what was to become the longest solo manned space flight in history, lasting about five days.

by the buoyant Soviet Union, and Titov was feted as a hero. Gagarin and Titov had also flown in what were to become the highest inclination orbits ever reached by humans, of 65°.

Another hero was American astronaut John Glenn, who finally got his country into orbit on 20 February 1962 aboard the Mercury capsule Friendship 7. Like the Russian dogs who had preceded Gagarin into orbit, Glenn had been preceded by the chimpanzee Enos in November 1961. Glenn's three orbits buoyed US morale and at last, it seemed, the nation had caught up with the Soviets. Another three-orbit mission was made by Scott Carpenter aboard Aurora 7 on 24 May 1962 and – despite Carpenter's error which resulted in missing the splashdown zone by 400km

It was a spectacular propaganda success for the Soviet Union, whose premier Nikita Khruschev had ordered the flight. Keen to perform further impressive feats, Khruschev ordered three cosmonauts to be launched into space because the US follow-up programme, Gemini, was to feature only two crewmen. The Soviets did not have a three-man spacecraft so a Vostok was reconfigured to take three crewmen crammed inside. As a result, the flight on 12 October of the craft re-named Voskhod 1 was the most risky in history.

THE VOSKHOD

The Voskhod spacecraft was basically a Vostok with some components removed and some added, increasing the weight to about 5.4 tonnes. A back-up retrorocket was required because launches on the uprated Vostok launcher, the SL-4 with a new upper stage, took the craft into a higher orbit which would not permit a natural decay within the ten-day deadline set for Vostok. The craft had a cup-shaped solid-propellant retrorocket added to the top of the spherical flight cabin.

The main difference from the Vostok was that the Voskhod had much of its interior removed so that up to three crewmen could lie side-by-side or, in the case of the future Voskhod 2, a flexible airlock could take the place of the third cosmonaut seat. This meant that the crew had no ejection seats – and no means to escape a launch failure. They also flew in tracksuits rather than spacesuits. With no ejection capability and having to land while still inside the capsule, the spacecraft needed to be fitted with a soft-landing retrorocket system. It was deployed under the main parachute and was fired just before touchdown, reducing the landing velocity to around 0.2 metres per second.

So, cosmonauts Vladimir Komarov – the only true pilot cosmonaut – a doctor, Boris Yegerov, and a space designer, Konstantin

Feoktistov, who had helped to reconfigure the craft, were launched successfully and came home after a flight of just one day. They did little during their mission because they were unable to move. However, the Soviet propaganda machine notched up another success, while the west assumed that a brand new spacecraft had been developed.

THE FIRST SPACE WALK

When the US Mercury programme finished, President John F. Kennedy made the bold decision in May 1961 to send astronauts to the Moon by 1969. A programme already planned was adapted to enable astronauts to gain all the experience necessary to make the Moon mission possible – long duration flight, manoeuvring the craft in space, rendezvous and docking with other spacecraft, and spacewalking, or extra-

Above: Astronaut Gordon Cooper demonstrates the cramped conditions inside the Mercury capsule during training for the flight in Faith 7.

A2 VOSKHOD

Right: The Voskhod launcher was uprated from the Vostok-launcher design to include a new upper stage.

vehicular activity (EVA). The project was called Gemini.

With NASA planning an EVA during a Gemini flight in 1965, the inevitable happened – a Soviet cosmonaut got there first. Alexei Leonov struggled to control his movements in space as he floated from the flexible airlock of Voskhod 2 on 18 March. After about 20 minutes cavorting in front of TV cameras showing pictures to the triumphant Soviet people, Leonov had to reduce the pressure of his spacesuit which had ballooned, so that he could squeeze back into the extendible flexible airlock. It was difficult and exhausting. Leonov and his commander, Pavel Belayev, returned home safely, after making a landing way off target, deep in a snowy forest after a retro-fire problem.

In Leonov's shadow, the Gemini programme began with a modest three-orbit test flight on 23 March 1965 by astronauts Gus Grissom – of Mercury fame – and John Young, a member of a new corps of nine astronauts appointed by NASA. The main achievements of the flight were the first manoeuvring in space, changing orbit and the use of a flight computer. The two-man, bell-shaped Gemini spacecraft came in two sections: a re-entry crew module and an adaptor section, which could be jettisoned, comprising mainly retrorockets. The complete spacecraft weighed about 3.25 tonnes and measured 5.58m (18ft 4in) long and 3.05m (10ft) in diameter at the base of the white adaptor section. The black-painted crew module was 3.35m (11ft) long and 2.28m (7ft 6in) in diameter at base. Each pilot had a small window, and the crew lay in ejection seats for launch escape capability. Each pilot had an entry hatch, easy enough to open manually to allow a crewman to exit making an EVA. The displays were similar to Mercury except that there was no periscope. The spacecraft was the first to be operated by an on-board computer.

SOVIET VOSKHOD MISSIONS

Date	Vehicle	Mission
6 October 1964	Cosmos 47	Successful unmanned test flight of Voskhod.
12 October 1964	Voskhod 1	Successful 1 day 00hr 17min flight of three cosmonauts, Vladimir Komarov, Konstantin Feoktistov and Boris Yegerov.
22 February 1965	Cosmos 57	Unmanned Voskhod 2 test flight explodes in orbit.
7 March 1965	Cosmos 59	Non-Voskhod test of EVA airlock.
18 March 1965	Voskhod 2	Flight lasts 1 day 2hr 2min and features the space walk, by Alexei Leonov, lasting about 20min. Commander was Pavel Belayev. Manual re-entry after systems failure.
22 February 1966	Cosmos 110	Dogs Veterok and Ugolok make 21 day orbital flight in Voskhod craft, and return safely to Earth.

In the nose of the spacecraft were the thrusters for the re-entry control system, which stabilized the vehicle for the retrorocket burn. The nose section also contained the parachute system. The parachute opened on a lanyard which pulled the Gemini into a horizontal position for landing, rather than base-first as was the case in Mercury. This posed a problem for the first crew which did not expect such a violent "yank" when the chute opened. The faceplate on Grissom's helmet was cracked as it slammed into the control panel. Future crews knew what to expect. Additional control thrusters for an orbital positioning and manoeuvring system were placed in the rear adaptor section which was split into two parts: one holding four retrorockets, totalling a 1.13 tonne thrust, and another carrying environmental supplies, such as oxygen and batteries for power. This latter section was deployed first, exposing the inner section for the retro burn. Once the retro fire had completed, the inner adaptor section was also jettisoned to prepare for the capsule's re-entry into the Earth's atmosphere.

Gemini craft carried rendezvous radar and associated equipment, and except for Gemini flights 3, 4, and 6, also carried unique oxygen-hydrogen fuel cells to generate electricity. A fuel cell changes chemical energy into electrical energy through the reaction between two chemicals, in this case liquid oxygen and hydrogen. A by-product of the reaction between oxygen and hydrogen is drinkable water which is consumed by the astronauts.

GEMINI TAKES THE LEAD

Gemini was launched by the US on its second generation ICBM, the Titan 2. There were two unmanned test flights before Gemini 3 flew. This mission was followed by nine more missions up to November 1966, and formed one of the most spectacular series of space flights in history. Each one made a step further towards the Moon by gaining all the experience required of the programme – and at no time was there any sign of a Russian cosmonaut. The Soviet bubble had burst.

Gemini 4 featured the first American EVA by Edward White on 3 June 1965 and a four-day mission, the longest yet by America. Gemini 5 extended this record to eight days after its launch on 21 August. Gemini 6 was supposed to perform the first rendezvous and docking with an Agena target vehicle, but which unfortunately blew up on 25 October.

After the first US Mercury programme ended, President John F. Kennedy made the bold decision in May 1961 to send astronauts to the Moon by 1969. The project was called Gemini.

VOSKHOD 2

Alexei Leonov became the first person to make a spacewalk on 18 March 1965. He exited from Voskhod 2 via an inflatable airlock which was extended from the crew cabin.

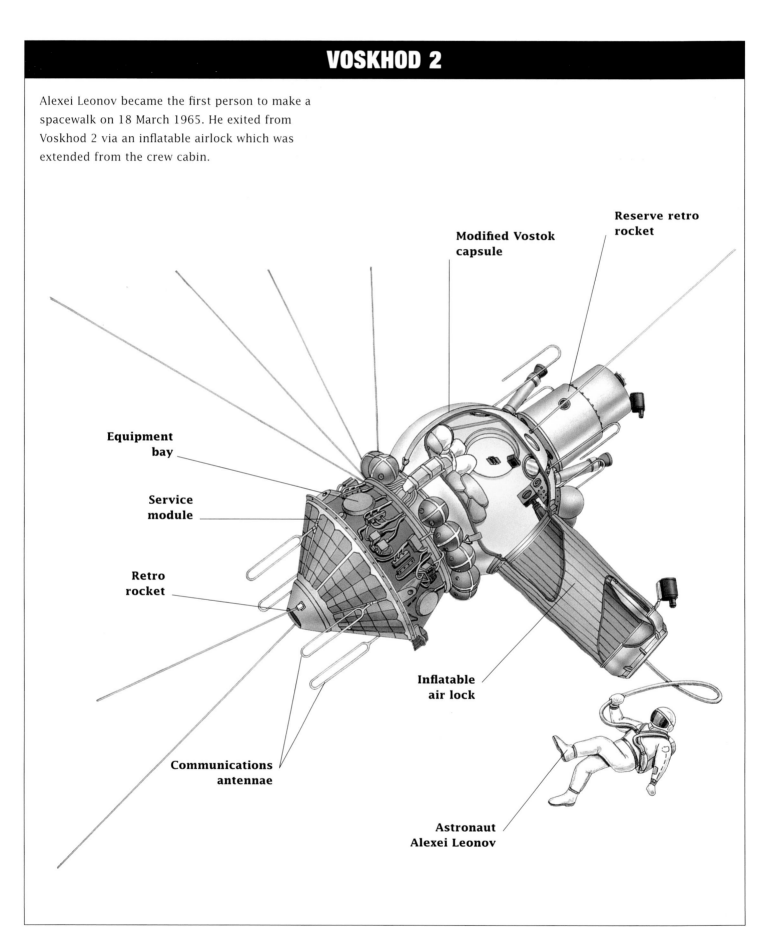

Reserve retro rocket

Modified Vostok capsule

Equipment bay

Service module

Retro rocket

Inflatable air lock

Communications antennae

Astronaut Alexei Leonov

Instead it was decided to fly Gemini 7 first in December, and then fly Gemini 6 to rendezvous with it. One of the most spectacular space missions was born. Frank Borman and James Lovell were launched in Gemini 7 on 4 December 1965, and extended the manned space flight record to almost 14 days. While in space they received a visit on 15 December from Gemini 6, piloted by Wally Schirra and Tom Stafford, which performed the first rendezvous, coming to within a foot of Gemini 7.

The next target for Gemini was an actual docking. This was achieved on 16 March 1966 by Neil Armstrong and David Scott who

US GEMINI MISSIONS

Date	Vehicle	Mission
8 April 1964	Gemini 1	Test flight of capsule attached to second stage of Titan launcher. No recovery planned.
19 January 1965	Gemini 2	After launch pad abort on 8 December 1964, flies to high altitude to test high-speed, high-heat re-entry and is recovered 18min after launch.
23 March 1965	Gemini 3	Three-orbit test of craft with astronauts Gus Grissom and John Young in 4hr 52min 51sec.
3 June 1965	Gemini 4	Flight of 4 days 1hr 56min under Commander Jim McDivitt; first US EVA by Ed White lasts 22min.
21 August 1965	Gemini 5	Record-breaking 7 day 22hr 55min flight by Gordon Cooper and Pete Conrad.
4 December 1965	Gemini 7	Record-breaking 13 day 18hr 35min flight by Frank Borman and James Lovell.
15 December 1965	Gemini 6	First rendezvous in space with Gemini 7 by Wally Schirra and Tom Stafford during 1 day 1hr 51min flight.
16 March 1966	Gemini 8	First docking in space with Agena target rocket by Neil Armstrong and David Scott. Mission then aborted by serious control problems. Lasts 10hr 41min.
3 June 1966	Gemini 9	Rendezvous with target craft but no docking. Record-breaking 2hr EVA by Gene Cernan. Commander is Tom Stafford. Mission lasts 3 days 0hr 20min.
18 July 1966	Gemini 10	Docking with Agena rocket which re-boosts Gemini to record 740km (462 mile) altitude. Crew is John Young and Michael Collins who made space walk. Mission lasts 2 days 22hr 46min.
12 September 1966	Gemini 11	Docking and reboost to record 1368km (855 mile) altitude, plus 44min space walk. Pete Conrad and Richard Gordon. Mission lasts 2 days 23hr 17min.
11 November 1966	Gemini 12	Docking with Agena, record EVA of over 2hr by Buzz Aldrin. Commander Jim Lovell. Mission lasts 3 days 22hr 31min.

Right: Gemini 1 is launched in March 1964, a year before the first manned Gemini flight.

GEMINI

Ten manned spaceflights aboard Gemini craft were made between 1965 and 1966, paving the way for Project Apollo, by achieving rendezvous, dockings, spacewalks and long duration flights.

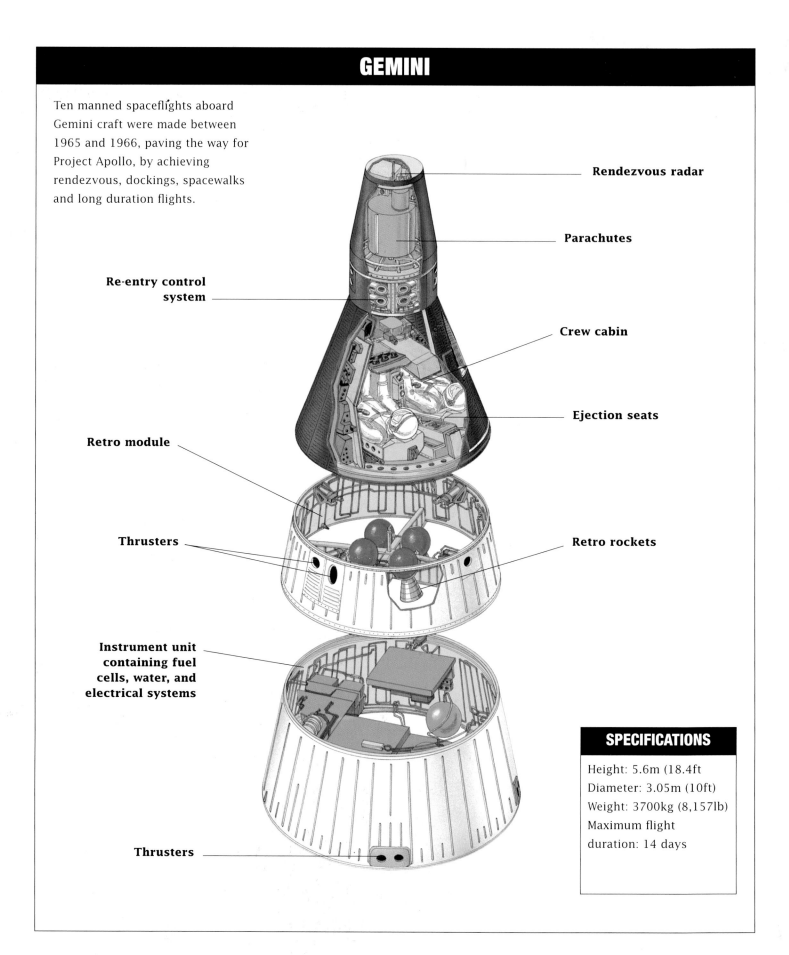

Rendezvous radar

Parachutes

Re-entry control system

Crew cabin

Ejection seats

Retro module

Thrusters

Retro rockets

Instrument unit containing fuel cells, water, and electrical systems

Thrusters

SPECIFICATIONS

Height: 5.6m (18.4ft
Diameter: 3.05m (10ft)
Weight: 3700kg (8,157lb)
Maximum flight
duration: 14 days

joined up with the Agena 8. However, the crew encountered problems and ended up spinning at one revolution per second due to a jammed thruster. In spite of the most dangerous moment yet in manned space flight, all the crew made it home safely. Gemini 9, launched on 3 June 1966, rendezvoused with a target rocket but was unable to dock because the payload shroud of the target had not detached. Pilot Gene Cernan also performed a record breaking two-hour space walk which created more problems than it solved. His "space walk from hell", as he described it later, was made almost impossible by the tendency to bounce off anything with which he made contact – a perfect illustration of Newton's law. Gemini 10 performed another Agena docking in July 1966. This time the rocket re-ignited its engine, boosting the crew to a higher altitude, a feat also achieved by Gemini 11 which in September 1966 reached a record altitude of 1368km (855 miles). Gemini 12 ended the programme with another

docking and a successful EVA, made possible by the use of handles and footholds on the outside of the spacecraft for the astronaut Buzz Aldrin.

Apollo and the moon beckoned.

Above: Gemini 6 is recovered from the Atlantic Ocean after its historic rendezvous mission.

SPINNING IN SPACE

Neil Armstrong and David Scott had just completed the first docking in space, linking their Gemini 8 spacecraft to the Agena 8 target vehicle on 16 March 1966 when they were sold a dummy. Mission control had noticed an anomaly on the Agena and told the crew that if they ran into any positional control problems on the Agena to turn off the control system. With this in mind, the crew went to their next task of commanding the Agena's system to turn 90° to the right. The manoeuvre took less than five seconds instead of the planned 60 seconds. Scott looked at the control panel and saw that the craft was in a 30° roll. Armstrong temporarily stopped the roll by firing Gemini's thrusters, but the coupled vehicles soon started to roll again. With the warning still ringing in their ears, the crew naturally blamed the Agena and shut off the target's control system. Armstrong prepared to move the vehicles into a correct horizontal position – then the real trouble started. The vehicle combination went into an accelerating roll. Still suspecting that the Agena was causing the roll, Armstrong struggled to isolate the problem in vain. Facing the danger of the combination breaking apart and causing a catastrophic collision, Armstrong decided to undock. Gemini's roll rate increased to one revolution per second and started to pitch as well. "We have serious problems here...we're tumbling end over end up here," Scott reported. "We're rolling up and can't turn anything off. Continuously increasing in a left roll," Armstrong added. The crew were getting dizzy, the dials on the control panel were blurred – and they were close to their physiological limits. The only way to bring the craft under control was to use the re-entry control system and disable the orbital manoeuvring system. Armstrong steadied the craft using the re-entry system, and then turned the orbital manoeuvring system on again, but the roll returned. At last, the crew realized what was going wrong – a thruster on the orbital manoeuvring system was still firing. Having used the vital re-entry system meant that the flight had to be aborted. Armstrong and Scott splashed down safely in the Pacific Ocean after their ordeal.

LUNAR LANDINGS

Even before the Space Age began, people had dreamed of flights to the Moon and beyond. Many scientists had designed spaceships, and imaginative drawings had been painted of what it would be like on the Moon when the first people landed. Nobody had seriously thought that people would travel to the Moon very early in the first phases of space exploration – but they did.

The US Project Apollo was an extraordinary programme, which now seems to have been conducted in a frenzy to get to the Moon as quickly as possible. The goal was not scientific so much as political, hardly surprising given the nature of the early days of the Space Age when an intense rivalry developed between the two superpowers. President John F. Kennedy made a commitment in May 1961 to land a man on the Moon by the end of 1969 – to beat the Soviet Union and ensure that technological leadership of the world lay in the west, not the east. It was a bold challenge. At the time, US experience of space amounted solely to Alan

Left: Charlie Duke explores the Descartes region of the Moon with the Lunar Roving Vehicle (LRV) during the Apollo 16 mission in April 1972.

Shepard's 15 minutes of manned space flight, and then only five minutes of that was in space. NASA scientists and engineers had to work quickly to discover the best way of reaching the Moon.

There were three methods by which NASA believed it could meet Kennedy's deadline, but not necessarily within budget. The first and most obvious way was a direct-ascent method. This would involve the construction of a huge booster, called the Nova, which would launch a large one-piece spacecraft directly to the Moon, land there and, after a short period of exploration, take off again and return to Earth. The Nova booster would generate up to 18 million kilogrammes (40 million pounds) of thrust. Although this was the most logical method, it was also the most expensive.

In an earth orbit rendezvous (EOR) method, all the components required for the lunar trip would be launched separately into an orbit of the Earth where they would

Right: The Apollo 11 spacecraft is pictured atop its Saturn V booster at the Kennedy Space Centre's Pad 39A. The launch escape system rocket is on the top of the spacecraft.

APOLLO FLIGHT PROFILE – LUNAR LANDING MISSION

This is the flight path of an Apollo mission to make a landing on the Moon and return to Earth, featuring Earth parking orbit, translunar coast, lunar orbit, lunar landing, and the return journey.

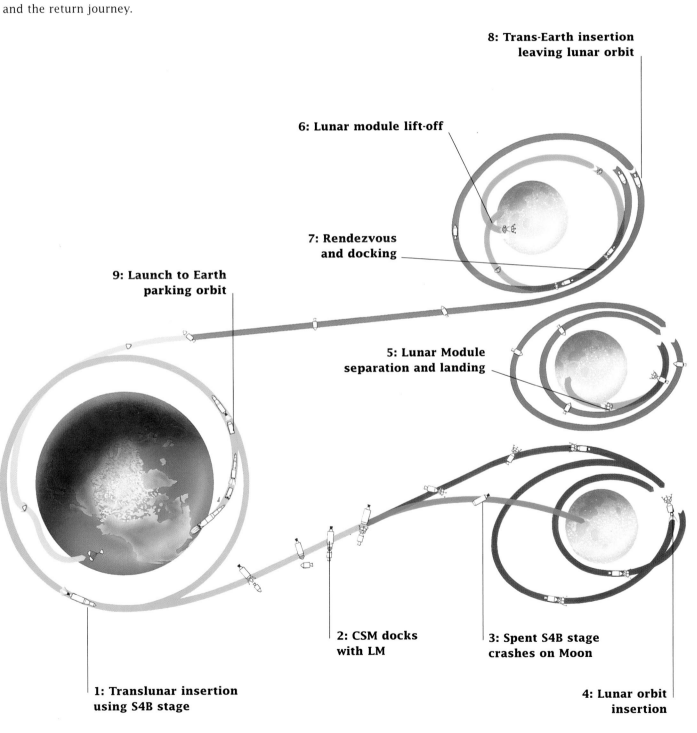

8: Trans-Earth insertion leaving lunar orbit

6: Lunar module lift-off

7: Rendezvous and docking

9: Launch to Earth parking orbit

5: Lunar Module separation and landing

2: CSM docks with LM

3: Spent S4B stage crashes on Moon

1: Translunar insertion using S4B stage

4: Lunar orbit insertion

Above: The Apollo 11 command/service module, Columbia, manned by Mike Collins as seen by the Eagle lunar module crew, Neil Armstrong and Buzz Aldrin, in lunar orbit.

Right: The ill-fated crew of Apollo 1, who perished in a launch pad fire are pictured during training. From left to right: Roger Chaffee, Edward White, and Gus Grissom.

remaining behind in the mother ship, the other two would man a landing vehicle and touch down on the moon using its descent engine. After a moon walk, the top half of the lunar excursion module (LEM) would take off, leaving the bottom half on the Moon, and redock with the mother ship in lunar orbit. The mother ship would then break out of lunar orbit and head back to Earth.

Since the rendezvous was taking place in lunar orbit there was no room for error. Some of the most difficult course corrections and manoeuvres had to be made after the spacecraft had been committed to a circumlunar flight. If the engine did not break the crew out of lunar orbit, they would be lost. The cheapest of the options available, the LOR method, was adopted by NASA in 1962, and project Apollo was on its way to meet its target within seven years. Project Gemini was to demonstrate the Apollo requirements in earth orbit, and unmanned Moon scouts were launched to find the best sites for a

rendezvous, be assembled into a single system, fuelled, and sent to the Moon. Smaller Saturn C-5 launch vehicles were already under development for the job. An attractive spin-off to the EOR approach was the establishment of a space station in earth orbit to serve as a base for the lunar mission's rendezvous, assembly and refuelling. The base could also serve scientific research purposes.

LUNAR ORBIT RENDEZVOUS

The prospect of constructing large components in orbit when a space rendezvous had not yet been accomplished alarmed NASA, and a third option was chosen: the single launch of a craft attached to a Saturn C-5 for a lunar orbit rendezvous (LOR) mission. It was the simplest of the three methods but also risky. Three astronauts would be launched in a mother ship, first reaching earth orbit, then heading for the Moon and entering an orbit around it. With one astronaut

LUNAR MODULE

This cutaway of the lunar module shows how the ascent stage used the descent stage as a launch pad on the Moon.

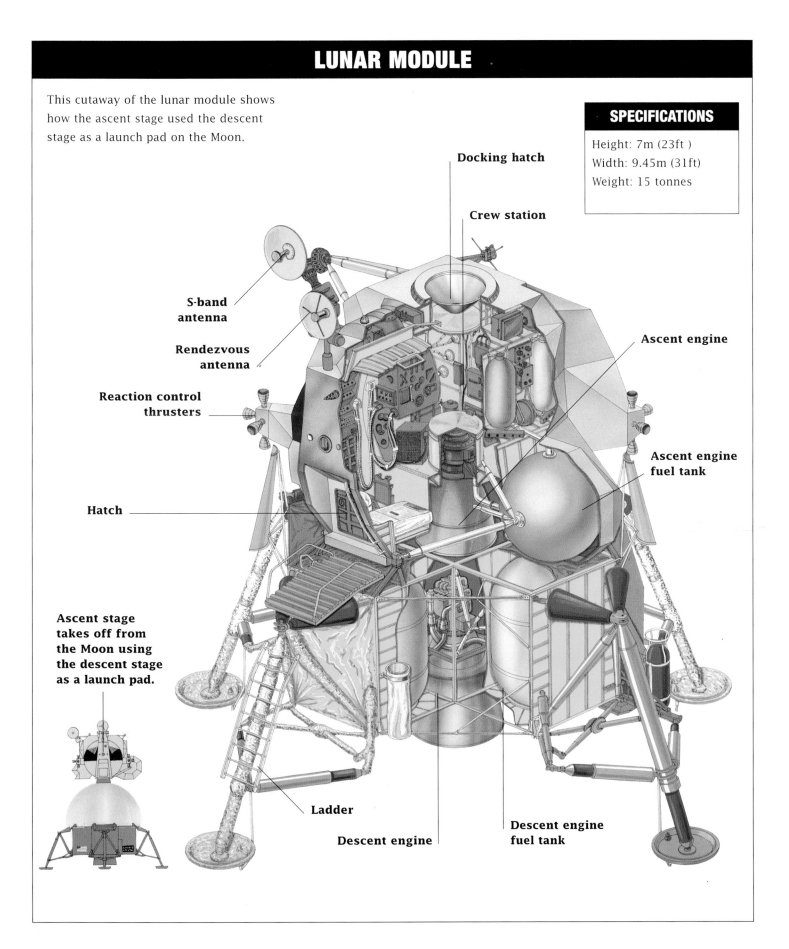

Docking hatch

Crew station

S-band antenna

Rendezvous antenna

Reaction control thrusters

Hatch

Ascent engine

Ascent engine fuel tank

Ascent stage takes off from the Moon using the descent stage as a launch pad.

Ladder

Descent engine

Descent engine fuel tank

> "12, 11, 10, 9, 8, ... ignition sequence start." A small ball of rich orange flame appeared at the base of the rocket, and then it suddenly exploded into a massive fire. "All engines running ..." The Saturn seemed to sit there for ever. "We have lift off!"

landing. Thus began an extraordinary period of space exploration in what was to become a significant decade, dubbed the "Swinging Sixties", a description in part due to the excitement of the rapidly developing lunar race.

THE APOLLO SPACECRAFT

The Apollo lunar spacecraft comprised three major components: the command module in which the crew of three were launched flying to and from the Moon; the service module, which was attached at all times to the command module until just before re-entry into the Earth's atmosphere; and the lunar module, in which two of the crew would land on the Moon and take off again to join the command module in lunar orbit. The command module (CM) and service module (SM) were known as the CSM to describe their combination, and the lunar module (LEM) was abbreviated to LM.

Positioned on top of the Apollo command module was a launch escape system (LES) which would be used during the first 100 seconds of the launch in the event of the Saturn V booster malfunctioning. The LES comprised a solid-propellant rocket with an all-over cover that protected the command module from the exhaust. If an abort was ordered, the LES rocket would fire, pulling the command module away from the rest of the rocket. It could then make a parachute drop into the sea, close to the Kennedy Space Centre

(KSC), a huge development north of Cape Canaveral dedicated to the lunar race. If not required, the escape system was jettisoned after 100 seconds, taking with it the boost-protective cover.

Apollo was assembled for the launch with the lunar module positioned underneath the command and service modules. Before Apollo could leave the Saturn V rocket and continue on its way to the moon, a transposition and docking manoeuvre had to take place. Once dispatched beyond Earth's orbit, the command and service module separated from the third stage of Saturn V, turned around and docked with the lunar module, using a docking drogue mechanism sited at the nose of the command module. The command module then extracted the lunar module from Saturn V, and the combined craft flew to the Moon. The crew were able to transfer between the two habitable modules via a tunnel, once the docking probe had been removed.

The five-tonne command module was about 3.65m (12ft) high and the same wide, and provided relatively luxurious space for the crew compared with the cramped Mercury and Gemini capsules. The command module served as the flight deck, sleeping quarters, kitchen, washroom and toilet, in which solid-waste bags were stored and a urine collection device was used to vent urine into space.

The crew sat in three reclining seats – one detachable – on the floor and faced a display console measuring a breadth of about 2.1m (7ft), with switches and dials for all the systems. These included the flight computer, manoeuvring thrusters and main service propulsion engine. In a small section at the foot of some couches, which also served as a private area for toileting, was the navigation bay. The atmospheric pressure in the cabin was 15 pounds per square inch, with 100 per cent oxygen. A vital part of the module was the ablative heat shield which protected the

SATURN V'S COMPUTER

Despite the brute force of the monster rocket, the Saturn V had what was then considered to be a highly sophisticated flight computer. Called the instrument unit, it was a ring-like structure mounted around the rocket which measured the booster's acceleration and directional position, and calculated the corrections necessary before commanding the engines' burn time. It also measured the booster's telemetry, electrical supply and thermal conditioning system. The unit was 0.9m (3ft) high, 6.4m (21ft) in diameter, and weighed over 45kg (99lb). It had the computing power of today's simplest pocket calculators.

THE CARNIVAL LAUNCH

On 9 November 1967, the first Saturn V booster was launched from Pad 39A at the Kennedy Space Centre, creating an extraordinary burst of noise, light and shock waves. At the press site, 4.8km (3 miles) away, the famous American TV anchor man Walter Cronkite reported on the launch as the roof of his studio began to fall in. Watching a Saturn launch was more of an event than a moment. For Apollo 15 on 26 July 1971, the sun rose to begin what was to be a typically sultry, perspiration-drenching Florida day. From the press site at the Kennedy Space Centre, looking about as high as a match stick held at arm's length, was the white needle, Saturn. The atmosphere at the press site was more like a carnival, with most people wearing cardboard sun-hats emblazoned with the names of soft drinks. Three men were about to be blasted away from Earth, maybe never to come home, and it was as if the press were reporting the Wimbledon tennis championships.

Eventually the moment arrived – the final countdown. Things became serious. The place went quiet as the voice of the NASA public affairs official read out the final seconds. Hundreds of cameras went into operation, with a sound like a multitude of crickets, as people heard, "12, 11, 10, 9, 8...ignition sequence start...". A small ball of rich orange flame appeared at the base of the rocket and then it suddenly exploded into a massive fire, spewing out clouds of steam and smoke either side of the launch pad. "All engines running...." The Saturn seemed to sit there for ever, churning away. "Ooooh, oooooh, aaah, ooooooh!" came the expressions from the crowd. "We have lift-off...." The Saturn was released. It rose slowly in an eerie silence. Seven seconds after lift-off that silence was broken. The ground trembled as a rumbling slowly seemed to engulf the press site. Then the noise arrived...a shattering cacophony of cracks of the sound waves and a roar like several low-flying jets on afterburner. It went on and on as the rocket flew higher, trailing a tongue of flame and smoke twice the rocket's length, and billowing clouds at least 300m (1000ft) into the air on each side of the pad. The noise reached its crescendo at T+25 seconds, began to diminish at T+40, and by T+60 was a gentle murmur in the sky. The experience was over. The launch pad was empty and steaming after its drenching from cooling water. Nobody spoke for minutes. The silent awestruck feeling was astonishing. People just stared. And soon laughs and giggles broke out again and the carnival atmosphere returned.

Right: The cataclysmic blast-off of Apollo 11 on 16 July 1969 emits a shattering roar, crackle and tremble that shakes the ground for miles around.

crew from temperatures reaching 3000°F, which would be experienced during the plunge into the Earth's atmosphere which began at a speed of about 40,000km/h (almost 25,000mph).

The service module, weighing 24 tonnes, was about 7.6m (25ft) long, with a conical rocket motor nozzle at its front end, and a large communications antenna on a short deployable boom. The module contained the 9.3 tonne-thrust service propulsion system engine, vital for the lunar orbit insertion and trans-Earth burns. In addition to the rocket's propellants, the service module contained fuel cells for electricity and water.

THE LUNAR MODULE

The lunar module looked like a strange insect; in fact it used to be called "the bug". It was a two-stage vehicle, 7m (23ft) high and 9.45m (31ft) wide, across its four spindly landing legs. Although it weighed about 15 tonnes, it was relatively fragile, being made of aluminium alloy with a thin layer of insulation. Indeed, the astronauts nicknamed it the "tissue-paper spacecraft".

The module came in two parts. The descent stage was unmanned and contained the descent engine to perform the lunar landing. To one of its legs was attached a ladder, down which the astronauts would climb to the Moon's surface. On top of the descent stage was the ascent stage, which was basically the flight deck and living quarters for the two crew members. There were no seats in the module; the crew stood at their flight consoles. The commander stood on the left and the lunar module pilot on the right, both able to look out through 0.3m (1ft) wide triangular windows. The main hatch opened inward and allowed the commander to go out first, crawling on his hands and knees onto a small porch on top of the ladder. The lunar module pilot was more of a systems engineer than a pilot.

Once the exploration was complete, the crew would then use the descent stage as a launch pad, and fire the ascent motor to fly

Below: The Apollo Saturn V booster was the tallest rocket ever developed and carried the first astronauts to the Moon.

SATURN V

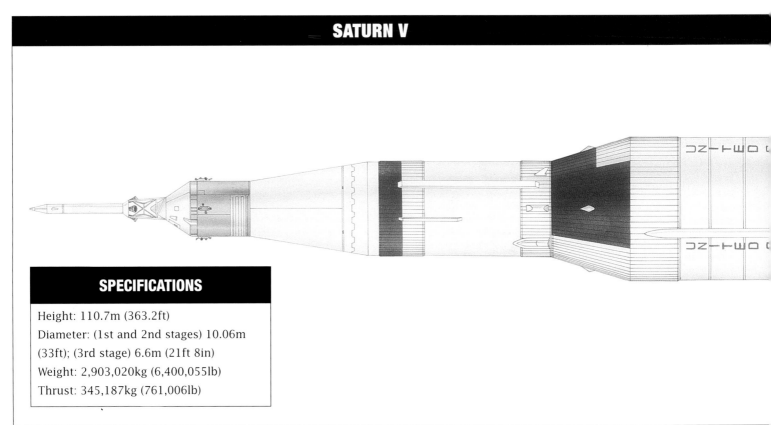

SPECIFICATIONS

Height: 110.7m (363.2ft)
Diameter: (1st and 2nd stages) 10.06m (33ft); (3rd stage) 6.6m (21ft 8in)
Weight: 2,903,020kg (6,400,055lb)
Thrust: 345,187kg (761,006lb)

Left: The Saturn V's third stage, the S4B was used to place Apollo in an Earth parking, and was fired again to despatch Apollo towards the Moon. The S4B was then discarded to impact the Moon or fly into solar orbit.

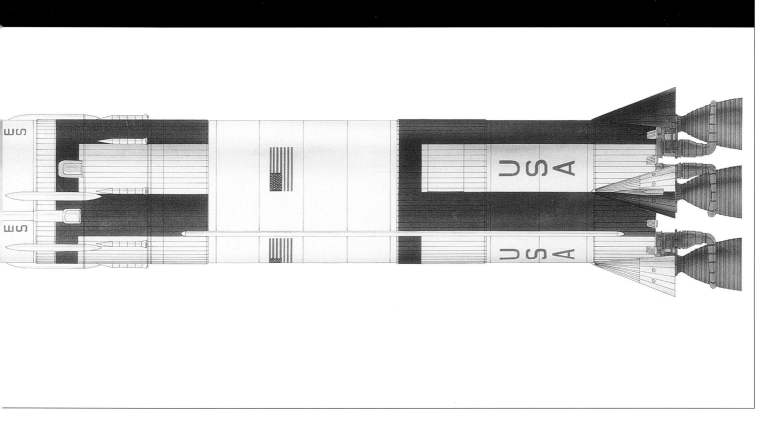

MANNED APOLLO FLIGHT LOG

11 October 1968 • APOLLO 7

Launched aboard a Saturn 1B booster into Earth orbit, the mission is commanded by Wally Schirra, with senior pilot Donn Eisele and pilot Walt Cunningham. A successful shakedown of the Apollo command and service modules in Earth orbit and lasts 10 days 20hr 9min 3sec.

21 December 1968 • APOLLO 8

One of the most historic voyages of exploration. Commanded by Frank Borman, with command module pilot James Lovell and pilot William Anders, the spacecraft carries the first humans to the Moon. The Apollo 8 crew makes 10 orbits around the Moon in 20hr 11min over a memorable Christmas on Earth, returning safely to Earth after a mission lasting 6 days 3hr 0min 42sec.

3 March 1969 • APOLLO 9

The lunar module is tested in space on a simulated Moon-landing and return within Earth orbit. Commander James McDivitt and lunar module pilot Rusty Schweickart flew the lunar module for 6hr 20min, later docking with command module pilot David Scott. The mission lasted 10 days 1hr 0min 54sec.

Above: The historic Earthrise image taken by Apollo 8 during the first manned orbital mission in December 1968.

18 May 1969 • APOLLO 10

Commander Tom Stafford, command-module pilot John Young and lunar-module pilot Gene Cernan are carried into lunar orbit in which Stafford and Cernan fly a simulated lunar landing, coming to within 14.5km (9 miles) of the Moon during an independent flight of 8hrs. The successful mission spends 2 days 13hr 31min in lunar orbit and lasts 8 days 0hr 3min 23sec in total.

16 July 1969 • APOLLO 11

The first attempt at a lunar landing is made by Neil Armstrong with command-module pilot Michael Collins and lunar-module pilot Edwin Aldrin. The lunar module Eagle lands safely on 20 July and later Armstrong sets foot upon the Moon on the Sea of Tranquillity. He and Aldrin spend 2hr 21min on the surface, collecting 22kg (48.5lb) of samples. Eagle stays on the Moon for 21hr 30min during an independent flight time of 1 day 3hr 59min. The command module, Columbia, is in lunar orbit for 2 days 11hr 30min and the total flight time is 8 days 3hr 18min 35sec.

14 November 1969 • APOLLO 12

The mission carries Charles "Pete" Conrad and Alan Bean to the Ocean of Storms where they land in lunar module Intrepid, while command-module pilot Richard Gordon remains in lunar orbit aboard Yankee Clipper. Conrad and Bean make two moonwalks lasting 7hr 45min, collecting 34kg (75lb) of samples. Independent flight time of Intrepid is 1 day 13hr 42min, including 1 day 7hr 31min on the surface. The total mission lasts 10 days 4hr 36min 25sec.

11 April 1970 • APOLLO 13

The ill-fated mission heads for the Fra Mauro highlands with commander James Lovell, command-module pilot Jack Swigert and lunar-module pilot Fred Haise. An explosion in the service module of Odyssey cripples the craft en route to the moon, preventing the planned landing by Aquarius.

The crew struggle home via a fly-by of the Moon during a drama that gripped the world. They splash down safely after a mission lasting 5 days 22hr 54min 41sec.

31 January 1971 • APOLLO 14

The mission that Apollo 13 is supposed to have accomplished now carries commander Alan Shepard (America's first man in space in 1961), command-module pilot Stuart Roosa and lunar-module pilot Edgar Mitchell on a journey lasting 9 days 0hr 2min 57sec. It includes the landing of Antares and the collection of 44.5kg (98lb) of samples during two moonwalks, lasting 9hr 22min, which feature the use of a wheeled tool and sample trolley. Shepard and Mitchell are on the surface for 1 day 9hr 31min out of Antares' independent flight time of 1 day 15hr 45min. The Kitty Hawk command module is in orbit for 2 days 18hr 39min.

26 July 1971 • APOLLO 15

Commander David Scott, command-module pilot Al Worden and lunar-module pilot James Irwin go into lunar orbit in which the command module Endeavour remains for 6 days 1hr 18min of the 12 day 7hr 11min 53sec mission. Falcon lands at Hadley Base, close to the Hadley Rille. Scott and Irwin operate the first lunar-roving vehicle during three moonwalks which last a total of 18hr 25min, during which 78.5kg (173lb) of samples are collected. The lunar stay lasts 2 days 18hr 55min. Worden makes a 38min EVA en route to Earth – the first of its kind. Probably the most scientifically successful mission of the whole programme.

16 April 1972 • APOLLO 16

The mission is commanded by John Young, with the Casper command-module pilot Ken Mattingly and the Orion lunar-module pilot Charlie Duke. Orion lands at Descartes and remains there for 2 days 23hr 14min, during which Young and Duke make three moonwalks lasting 20hr 14min, also featuring a lunar rover, collecting 96.6kg (213lb) of samples. Orion's independent flight time is 3 days 9hr 28min. Of Casper's 5 days 5hr 53min in lunar orbit, Mattingly spends 3 days 9hr 28min on his own – the longest solo US space flight. Total mission time is 11 days 1hr 51min 5s. Mattingly performs a 1hr 13min EVA on the journey home.

Below: Astronaut Jim Irwin pictured next to the first Lunar Roving Vehicle at Hadley Base during the Apollo 15 mission.

7 December 1972 • APOLLO 17

The final mission features a spectacular landing site at Taurus Littrow and the flight of the first astronaut-geologist, Jack Schmitt, the lunar-module pilot. The commander is Gene Cernan and command-module pilot Ron Evans. The lunar module Challenger is on the surface for 3 days 2hr 59min. During three moonwalks, lasting a total of 22hr 5min, including drives of a lunar rover, Cernan and Schmitt collect 110kg (243kg) of samples. The second moon walk lasting 7hr 37min is the longest in the history of the programme. Challenger ends its independent flight time of 3 days 8hr 10min by docking with the command module America, which is in lunar orbit for 6 days 3hr 48min of the entire mission lasting 12 days 13hr 51min 59sec. Evans makes a 1hr 6min EVA on the way home.

into lunar orbit for rendezvous and docking with the CSM, manned by the lone command module pilot. The service propulsion system would then be fired to take the crew out of lunar orbit, en route for Earth. The aiming point for earth was absolutely critical, with very little margin for error. As the CSM/LM plummeted towards Earth, the command module would have to separate and hit the atmosphere like a stone skipping across water. This operation basically slowed the capsule down sufficiently to make its final plunge into the atmosphere. Re-entry was controlled by slightly rolling the craft to ensure an accurate landing in the Pacific Ocean. If the command module hit the atmosphere too directly, at anything close to 90°, it would be crushed, burned and destroyed in seconds; and if the craft hit at an angle too acute, it would bounce out into space, never to return.

By 1968, versions of the Apollo modules had made a number of testflights in earth

Above: Buzz Aldrin descends the ladder of the Apollo 11 lunar module.

Right: Apollo 12's Alan Bean hard at work at the side of the lunar module Intrepid, as he prepares to deploy more instruments on the surface of the Ocean of Storms.

LUNAR ROVING VEHICLE

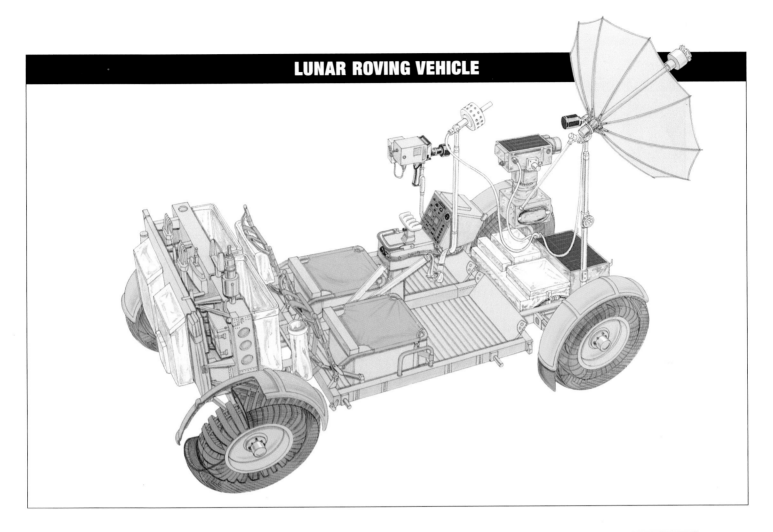

orbit using smaller Saturn 1 and 1B boosters. The mighty Saturn V, also flight-tested, was a gigantic booster and the star of the show, and built specifically to land men on the Moon. The Saturn V was designed and developed by NASA's Marshall Space Flight Centre at Hunstville, Alabama, under Dr Wernher von Braun, the former German V2 rocket engineer. Coupled with the Apollo system built on top, the monster rocket was 110m (361ft) high. Its five first-stage engines generated a thrust of 3400 tonnes, and each engine consumed almost 136 tonnes of propellant every second, creating an extraordinary noise, which shook the ground for miles around like an earthquake. The rocket consisted of three stages, two of which had served their purpose within nine minutes.

The first stage (F-1) was powered by a group of liquid-oxygen and kerosene engines which carried the vehicle to an altitude of 60km (37 miles) in 160 seconds, at a speed of 13,360km per second (8,300mph). When this stage was jettisoned, the SII second stage was then ignited. This was powered by five J2 engines which consumed liquid oxygen and liquid hydrogen, cryogenic propellants, for six minutes 30 seconds, by which time the rocket was at an altitude of 182km (113 miles), and moving at about 24,480km per second (15,200mph). The third stage was the re-ignitable SIVB, also powered by one cryogenic J2 engine. This burned for about two and a half minutes, achieving Earth orbital speed. This stage, with its Apollo payload, then circled the Earth in a "parking orbit" until the time scheduled to ignite the engine again, setting the Apollo mission on course for the Moon on what was called the trans-lunar injection.

Above: The Lunar Roving Vehicle (LRV) greatly extended the area of the lunar surface which could be explored by astronauts of the Apollo 15, 16, and 17 missions. Deceptive in appearance it looked like a simple "dune buggy" but in fact was a specialized space vehicle designed to operate in conditions of vacuum, wide extremes of temperature and over difficult terrain.

ZOND AND SOYUZ

The Soviet Zond lunar spacecraft was based on the design of the manned Soyuz Earth-orbital spacecraft.

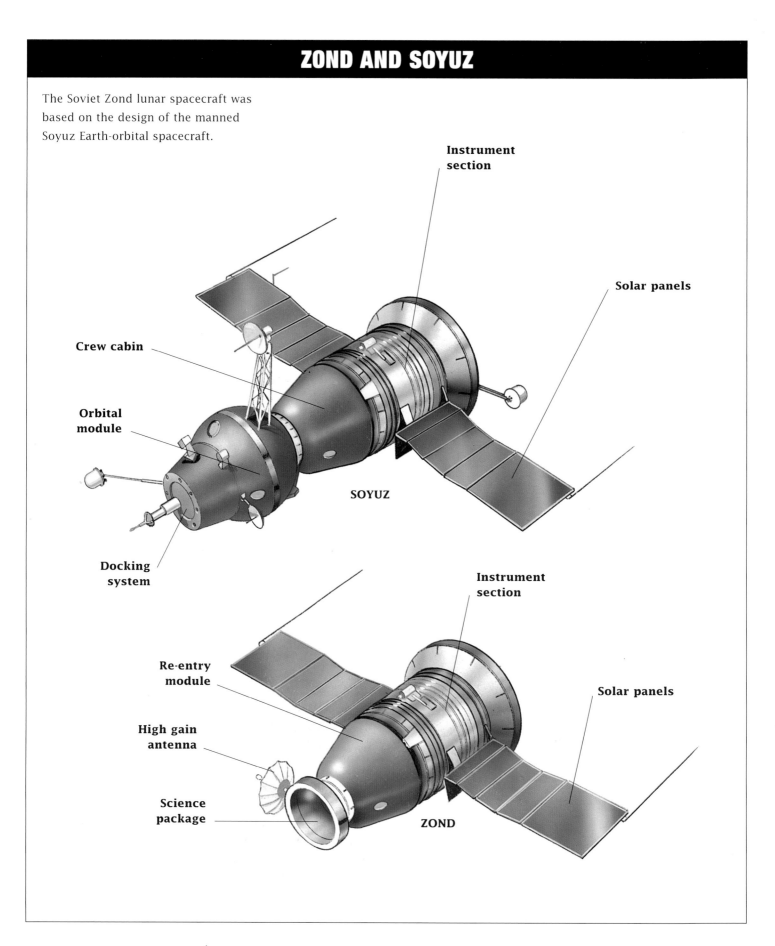

Instrument section

Solar panels

Crew cabin

Orbital module

Docking system

SOYUZ

Instrument section

Solar panels

Re-entry module

High gain antenna

Science package

ZOND

The single J2 fired for about five minutes, increasing the velocity to 39,040km per second (24,260mph), which allowed the Apollo crew to escape the pull of earth's gravity (escape velocity). After the lunar module had been extracted from the nose of the S4B, the spent stage was discarded and sailed into deep space, possibly entering solar orbit or even impacting on the Moon.

SOYUZ AND ZOND

By 1968, after a series of varied test flights, all was ready for the Apollo manned flights to begin, aiming for a moon landing in 1969. It was also assumed, of course, that the Soviet Union fully intended to beat America to the Moon. The Soviets had two plans in place to steal the US's thunder, first by flying cosmonauts around the Moon, and second, by landing a lone cosmonaut on the Moon before Apollo. The Soviets developed two new vehicles, a Soyuz manned transporter, and a giant rocket, called the N1. The Soyuz was a much more sophisticated vehicle than the earlier Vostok/Voskhod craft. Designed to carry two or three cosmonauts, the Soyuz was able to rendezvous and dock in space, and became a vital component for a planned space-station programme. It consisted of an orbital module with a docking system, flight cabin and re-entry vehicle, and a service module equipped with solar panels to provide on-board electrical power, with its own engine for manoeuvring. In a slightly modified form, the service module was also designed to be capable of undertaking trips to and from the Moon.

The Soyuz, without an orbital module, was designed to be launched as a Zond spacecraft on figure-of-eight round-trips to the Moon. It was not intended to orbit the moon. The Zond would fly unmanned missions first, and then be manned by two cosmonauts. The craft was planned to be launched on a relatively new Proton booster. This programme was called the L-1. The first prototype Soyuz was tested in earth orbit

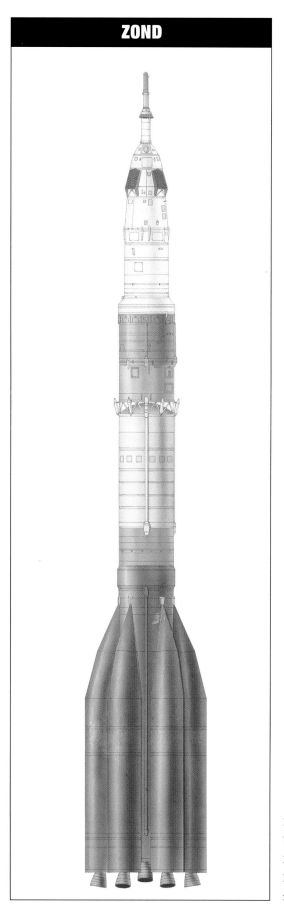

ZOND

Left: **The modified Proton booster which carried the Zond spacecraft. It was launched 11 times, with four failures.**

LUNAR ZOND FLIGHT LOG

Date	Flight Log
10 March 1967	Cosmos 146 makes test flight in Earth orbit.
8 April 1967	Cosmos 154 fails to reach correct Earth orbit.
28 September 1967	Launch fails.
22 November 1967	Launch fails.
2 March 1968	Zond 4 flies around the Moon but is deliberately destroyed as it is heading for a landing outside Soviet control.
23 April 1968	Launch fails.
14 September 1968	Zond 5 flies around the Moon and lands in the Indian Ocean after a control failure places the craft on such a trajectory into the Earth's atmosphere that a crew would have been killed.
10 November 1968	Zond 6 flies around the Moon but is depressurized during the return to Earth. The parachute fails and the craft is destroyed. Both malfunctions would have killed a crew.
20 January 1969	Launch fails.
8 August 1969	Zond 7 is the only completely successful mission, looping the Moon and returning safely to the Soviet Union.
20 October 1970	Zond 8 makes a lunar loop but control system failure results in landing in the Indian Ocean.

Note: Zonds 1–3 are failed probes to Venus and Mars.

Right: The lift-off of an N1 booster from the Baikonur Cosmodrome in Kazakhstan. All four flights of the Soviet mega-booster failed.

and unmanned in November 1966. Although the Zond did indeed loop the Moon and return to Earth, prompting the American reaction to fly an Apollo crew into lunar orbit in late 1968, the Zond was never manned. Indeed, had it flown with a crew on several of the unmanned missions, the cosmonauts would have been killed as a result of the malfunctions that plagued the craft.

THE SOVIET LUNAR PROGRAMME

The Soviet lunar landing programme was called the L-3. If Apollo is now regarded as extremely ambitious, then the L-3 was simply dangerous. It would be based on Soyuz-Zond technology and the giant N1, the Soviet's answer to the Saturn. The N1 was an extraordinary-looking machine. It was an almost uniformly tapered vehicle with a very long payload shroud at its top, in which the manned Moon landing craft was placed. The N1 had a base diameter of about

LUNA 1 LAUNCH VEHICLE THIRD STAGE ASSEMBLY

A modification of this upper stage was later used for the manned Vostok flights, for which the engine performance was improved and a protective cylindrical shroud which covered the Vostok retro-rocket was added to the rocket stage. It was also used for the occasional launch of unmanned non-recoverable satellites in the Soviet Earth resources programme.

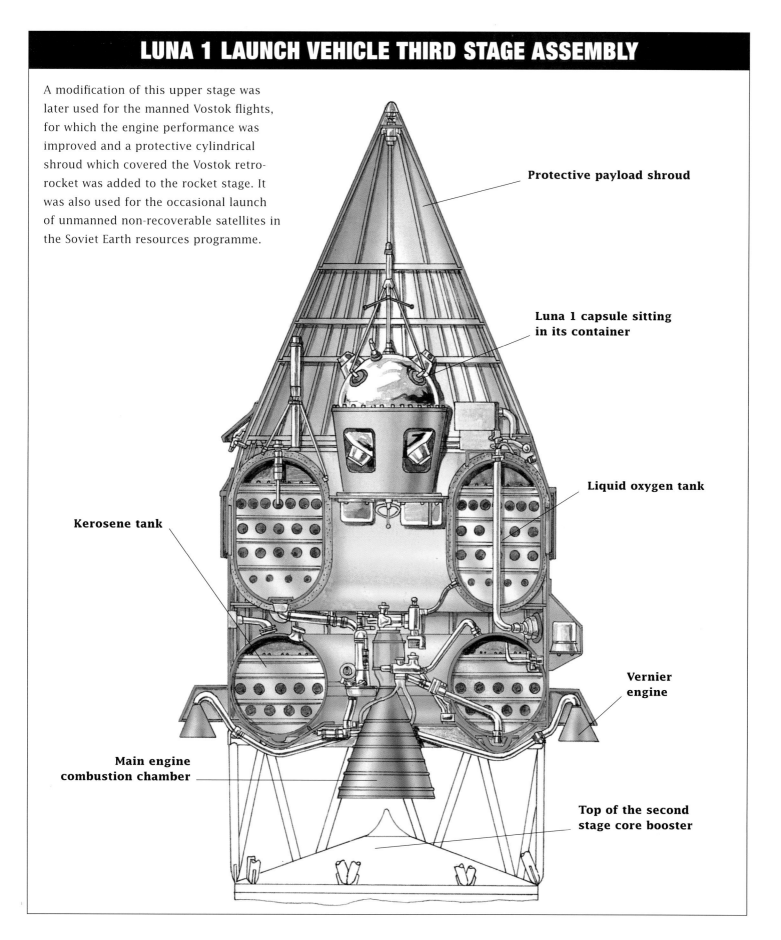

Protective payload shroud

Luna 1 capsule sitting in its container

Liquid oxygen tank

Kerosene tank

Vernier engine

Main engine combustion chamber

Top of the second stage core booster

Right: The entire N1 booster stack is rolled out to the launch pad to be erected into the upright position by a special crane structure.

SOVIET N1 LAUNCH ATTEMPTS

26 February 1969

The first N1 giant booster, carrying a simulated Moon lander rises from its Baikonur launch pad and all seems to be going well. At T+66sec, an oxidizer pipe to one of the 30 first-stage engines ruptures and leaks liquid oxygen which then catches fire. The booster continues for four more seconds but the engines are shut down by the flight computer and the launch escape system fires, carrying the lander to safety before an enormous conflagration ends the mission.

3 July 1969

The second N1 carries another simulated Moon lander, and almost as soon as it has risen from the launch pad a metallic object falls into the oxidizer pump of engine No 8, which explodes, disabling other engines and control cables. The vehicle falls back onto the pad and explodes, destroying not only its own launch pad but an adjoining N1 launch complex.

27 June 1971

N1–3 lifts off from the repaired adjoining launch pad, carrying a mock-up of the entire Moon lander system and almost immediately experiences roll problems. By T+39sec the roll has exceeded the limits of the launcher's control system. At T+48sec, the N1's second stage starts to break apart and at T+51sec the automatic flight control system shuts down all the engines, and yet another N1 plunges into the Baikonur steppe.

23 November 1972

The final N1 is loaded with a similar payload to N1-3, and at T+90sec, the planned shutdown of central engines of the N1 first stage causes an overload of pressure which ruptures propellant lines, and the vehicle catches fire. About 20 seconds later the first stage explodes.

By December 1972, 12 men had walked on the Moon. Today the Apollo programme seems like an extraordinary indulgence of technology, and it is unlikely that mankind will return to the Moon for at least another 30 years.

15.25m (50ft) and was just over 91m (300ft) tall. The 24.3m (80ft) long first stage was equipped with 30 liquid-oxygen and kerosene NK-33 engines with a thrust of around 5000 tonnes. The engines would operate for one minute 50 seconds, then the second stage, powered by eight similar NK-34 engines, would take over and operate for two minutes 10 seconds. The third and final stage of the booster itself was powered by NK-39 engines, which fired for six minutes 40 seconds. By this time the L-3 manned spacecraft would have reached earth parking orbit.

The two-man L-3 craft was made up of four parts: two rocket stages, a lunar orbiter and a lunar lander. The first rocket motor, called a Block G, would send the rocket combination on a flight towards the Moon and then be discarded. The next motor, called the Block D, was fired to place the lunarcraft in orbit around the Moon with an eventual low point of around 16km (10 miles). The plan to land a man on the lunar surface then took an unlikely turn. One cosmonaut would put on a spacesuit and make a space walk to transfer externally rather than internally to the lunar landing craft below. His companion remained in the mother ship, based largely on a Soyuz capsule, which would then separate from the Block D and landing craft combination.

The Block D engine would fire for the descent to the lunar surface. About 2.4km

LUNA 16

Launched in September 1970, Luna 16 was the first automatic spacecraft to voyage to the Moon, collect samples of soil and rock by remote control, and return them successfully to Earth. The principal piece of equipment carried was the soil sampler which consisted of an electric drill controlled from Earth. The Luna 16 project was followed by four more similar missions, two of which were completed successfully.

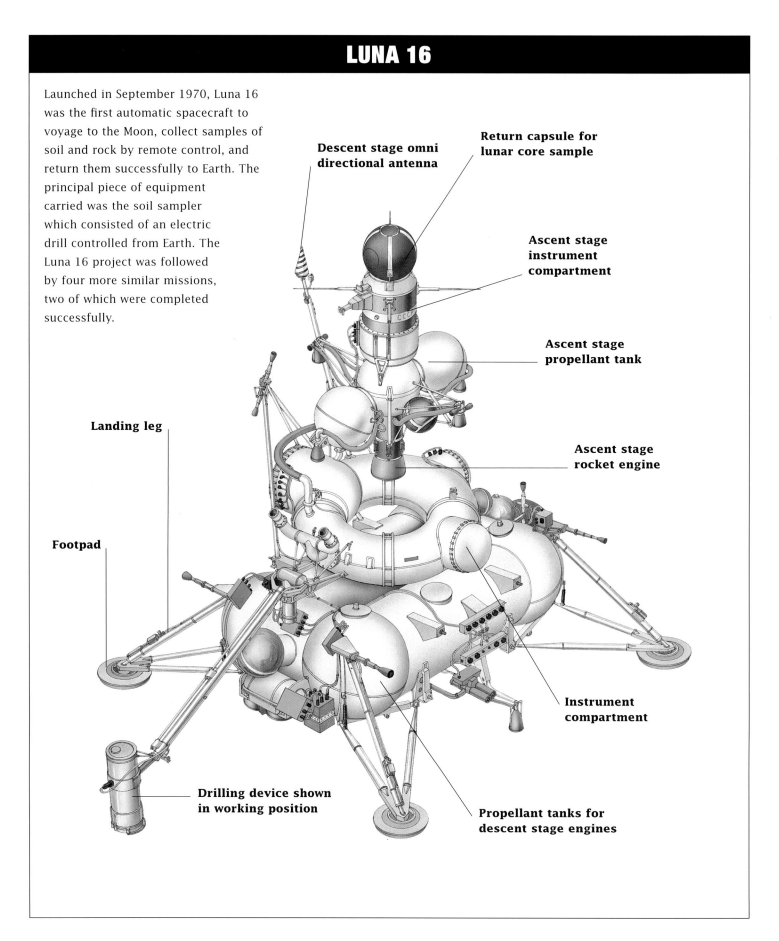

Descent stage omni directional antenna

Return capsule for lunar core sample

Ascent stage instrument compartment

Ascent stage propellant tank

Landing leg

Ascent stage rocket engine

Footpad

Instrument compartment

Drilling device shown in working position

Propellant tanks for descent stage engines

Left: One Soviet cosmonaut was to have landed on the Moon and walked upon its surface, flying this lunar module spacecraft equipped with just one engine.

(1.5 miles) above the surface, the 4.47m (14.5ft) high lander would separate and make a soft landing using its own small engine, while the Block D crashed nearby. The lone cosmonaut would exit via a circular hatch down a ladder onto the lunar surface, planting the Hammer and Sickle flag. After just an hour, he would return to the cabin with samples of lunar rock. Unlike Apollo, the entire lunar module would take off from the Moon, using the same descent engine, and dock with the lunar orbiter. The landing-craft cosmonaut would transfer back to the orbiter during another EVA, grasping his samples and moonrock for all he is worth. The lunar lander would be jettisoned, and the lunar orbiter would fire its engine and head for home in a similar profile to that used by the L-1 missions.

These plans depended on the successful testing of the Soyuz spacecraft and the N1 booster, but these tests did not go according to plan. In the final event, the US found itself racing alone to the Moon. By December 1972, 12 men had walked on the lunar surface, returning a total of 385kg (849lb) of Moon dust to Earth. Kennedy's goal had been achieved. However, once that target was reached, the Apollo programme quickly faded away, and other planned moon missions were cancelled. In hindsight, Apollo seems an extraordinary indulgence of technology. Nonetheless, for those who witnessed it, the first walk on the Moon will always remain as one of the most exciting events of the 20th century. It is unlikely that mankind will return to the lunar surface for at least another 30 years or more.

LAUNCH VEHICLES

Launch vehicles are required to send satellites into a variety of orbits which are prescribed in terms of altitude and inclination to the Earth's equator. A satellite has to travel at least 28,900km per hour (18,000mph) to enter a low Earth orbit of, for example, about 200km (124 miles). It will take the satellite about one and a half hours to make one revolution of Earth. This is called the orbital period. The higher the satellite's orbit, the longer it takes to travel around Earth.

Not all orbits are circular. A satellite may be in an orbit 200km (124 miles) at its low point, called the perigee, and 600km (373 miles) at its high point, the apogee. The perigee, apogee, inclination and period of orbits are important parameters for different types of satellites, which have to be met perfectly by the performance of the launchers. They must fly at the precise angle and direction, called azimuth, and in the appropriate speed increments to reach the correct orbit and position above Earth at the projected time.

Left: The US Delta III unmanned satellite ready for launch on Pad No. 17 at Cape Canaveral in 1998.

An X-ray astronomical telescope may be placed in an orbit between 114,000km (70,836 miles) and 6880km (4275 miles), and will take 48 hours to make one revolution of the Earth, enabling the craft to locate far beyond the Earth's interfering radiation belts for as long as possible. At an inclination of 40° to the equator during each orbit, the telescope will be within range of two ground stations in Australia and South America for an uninterrupted 40 hours.

An imaging spy satellite or a remote-sensing Earth-observation satellite will enter a low orbit, flying over the poles of the Earth. A satellite flying over the north and south pole at a height of about 400km (249 miles) will make 17 orbits while the Earth rotates once beneath it in a day. The polar orbit also has to be one that is "sun synchronous", meaning that every 24 hours a satellite passes over the same place on the equator at the same time of day. This makes it possible for a series of images to be taken of the same area at a similar sun angle, thereby noting any changes that have occurred in the area, such as the extent of flooding, since the previous images were taken. It is this facility to compare data which is so vital in the remote-sensing industry.

A communications satellite has an altogether different type of orbit. If a satellite can be placed in a circular orbit that goes around the equator, at 0° inclination, at such a height that it travels at the same speed as earth rotates, it will appear as stationary in the sky. That altitude is about 36,000km (22,369 miles). The orbit is called geostationary (GEO) and is populated by hundreds of communications satellites, such as those that relay TV images direct to people's homes. The rocket launcher that carries the communications satellite usually places its upper stage and satellite payload in a GEO transfer orbit with a low perigee and GEO apogee. Then a motor on the satellite fires, placing the satellite in GEO.

The closer a launch site is located to the equator – at 0° – the better it will be for GEO launches, because savings in rocket propellant can mean a larger payload may be carried. The further south or north a rocket is launched, the more energy it will require to reach an equatorial orbit. This explains why one of the most favourable launch sites for communications satellites is at Kourou, Guiana in South America, close to the equator. The International Sea Launch Company has gone one step further by actually putting a launch pad on the equator. The Sea Launch Odyssey launch pad is based on a semi-submersible oil rig, and is floated into position with its Russian-based Zenit 3L rocket.

As for polar orbiting satellites, some launch sites would be too dangerous to use. It is risky to launch into polar orbit from Cape Canaveral in Florida, as a rocket would have to fly over land. From Vandenberg in California, however, a rocket can be launched straight "down" the Pacific towards the south pole. Russia launches its rockets over vast areas of uninhabited land but nonetheless nomadic populations, such as those in the Kazakhstan steppes, have been known to use crashed spent rocket stages as shelters, while others have been affected by propellant showering down on the land after launch-failure explosions. Cape Canaveral is used for launches that can be made safely over the Atlantic in low-inclination orbits such as 37°.

Satellites have also been launched from the air and from under the sea. The US Pegasus satellite launcher is a winged three or four-stage rocket that is released from

Below: The sprawling Ariane launch base at Kourou, French Guiana, which supports launches of Ariane 4 and 5 boosters by the commercial organisation Arianespace.

Above: A winged Pegasus satellite launcher rides under the wing of a Boeing B-52 carrier aircraft to be dropped at high altitude to begin a commercial launch.

Right: A spectacular lift-off of a Soyuz booster operated by the commercial organization Starsem, carrying communications satellites into low Earth orbit.

under a carrier aircraft, thereby gaining thousands of feet in altitude without having to fire its own engines. Once released, the solid-propellant rocket engine of the first stage fires and begins to climb to orbit. The Pegasus can carry about half a tonne into a low Earth orbit. A Russian organisation is now marketing launches using one of many models of former Russian military missiles. Called the Shtil 2, the rocket was fired from a submarine in the Barents Sea for the first time in 1998, carrying a small satellite into orbit for Germany.

HISTORIC LAUNCH SITES

The first satellite, Sputnik 1, was launched from a remote launch site in Kazakhstan east of the Aral Sea. It became known as the Baikonur Cosmodrome but was actually 400km (249 miles) away from the Soviet town of that name. The cosmodrome is located close to the original small town of Tyuratam which today is a railway junction for the Moscow to Tashkent line. The locals say that Tyuratam is named after the

"burial place of the arrows", or "the burial place of Tyura", a son of the infamous Ghengis Khan who was killed in battle close by. In 1955, when Tyuratam was the location of open-cast mines on the Kazakh steppes, a 28km (17.4 miles) spur from the main railway line was laid, leading to one of the open-mine pits. A concrete launching pad was built here and became the site for launching the Soviet ICBM. The same Pad No.1 was the site from where Sputnik 1 and Yuri Gagarin departed, and is still in use for Soyuz launches. The site was named Baikonur in 1961 by the Russians who needed to identify the origination point of Gagarin's epic flight so that it could be ratified. Uninspired Soviet officials searched for an appropriate name for such an historic site and ended up choosing the name of the distant town of Baikonur because it meant "rich region".

The US's "Baikonur", Cape Canaveral, had an equally interesting history. The name Cape Canaveral is synonymous with the Space Age but is hardly the place one would associate with space. It started life

Today, Cape Canaveral in Florida is a busy Air Force base where unmanned satellites are launched from the same pads used in the heady days of the 1960s, when the Cape was strewn with lines of gantries along every shore.

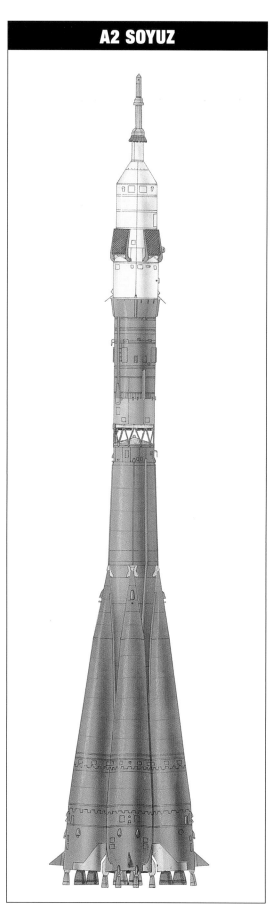

A2 SOYUZ

Right: The Soyuz booster was based on the original Soviet ICBM, and was equipped with an improved upper stage and a launch escape system rocket.

as a lush sub-tropical peninsula of sand jutting out into the Atlantic Ocean, half way down the eastern side of Florida. It was a desolate expanse of water, sand, gnarled brush, palmetto scrub and palm trees, inhabited by many species of wildlife, including alligators, snakes, sub-tropical birds and hordes of voracious mosquitoes. When Juan de Leon discovered Florida in 1509, the Cape was inhabited by native Americans. It remained almost totally undisturbed through three centuries of Spanish, French and British rule until just after the American Civil War, when the town of Cocoa was established south-west of the Cape and a lighthouse built on the giant sand spit. The lighthouse remains there today, often mistaken by observers for a rocket about to blast off. One embarrassed press photographer took photos of it while the real rocket blasted away elsewhere! Construction of a blockhouse and launch pad on the Cape began in 1950, and the first rocket was launched from there on 24 July the same year.

Today the Cape is a busy Air Force base where unmanned satellites are still launched from the same pads used in the heady days of the 1960s, when the Cape was strewn with lines of gantries along every shore. Many of these gantries have been demolished, including Pad 14 from where John Glenn was launched to make the first US manned orbital flight. North of the Cape is the NASA Kennedy Space Centre (KSC), from where the Space Shuttle is now operated. The KSC was developed to support the Apollo Moon programme and much of its infrastructure is used for the Shuttle, including the huge Vehicle Assembly Building and the launch pads.

FROM THE COLD WAR TO DETENTE

After more than 40 years into the Space Age, it is a surprising fact that, with thousands of launches under their belts, the four workhorses of the US and Russian launch industry – Delta, Atlas, Titan and

Left: An Atlas Agena rocket, carrying Ranger 7, is launched from Cape Canaveral in 1964.

ORIGINATION OF US LAUNCHERS 1958-1967

Over the first 10 years of the Space Age, the workhorses of the US launch industry were all based on a first stage that was, in effect, a ballistic missile.

First successful launch	Vehicle	Ancestor missile
January 1958	Jupiter C	Redstone IRBM
October 1958	Thor Able	IRBM, with upper stage
December 1958	Juno II	Jupiter missile with upper stage
December 1958	Atlas B	ICBM
February 1959	Thor Agena A	IRBM with upper stage
April 1960	Thor Able Star	IRBM with upper stage
May 1960	Atlas Agena A	ICBM with upper stage
October 1960	Thor Agena B	Thor Agena A
November 1960	Delta	Thor Able
May 1961 (manned)	Redstone	IRBM
July 1961	Atlas Agena B	Atlas Agena A
September 1961	Atlas D	ICBM
Jun 1962	Thor Agena D	Thor Agena B
May 1963	TAT-Agena	Thor Agena D with solid rockets
June 1963	TAT-Agena B	TAT-Agena D
July 1963	Atlas Agena D	Atlas Agena B
November 1963	Atlas Centaur	Atlas D with upper stage
April 1964	Titan II	ICBM 2nd generation
August 1964	TAD	Delta with solid rockets
December 1964	Titan IIIA	Titan I
January 1965	Thor Altair	Thor with different stage
June 1965	Titan IIIC	Titan IIIA
July 1966	Titan IIIB-Agena D	Titan IIIA
August 1966	Thorad Agena D	Thor Agena D
September 1966	Thor Burner II	Thor with different stage
May 1967	LTTAT Agena D	Long tank TAT Agena

Right: An uprated Atlas ICBM booster, called the Atlas E/F, was also used to launch satellites.

THRUST AUGMENTED DELTA

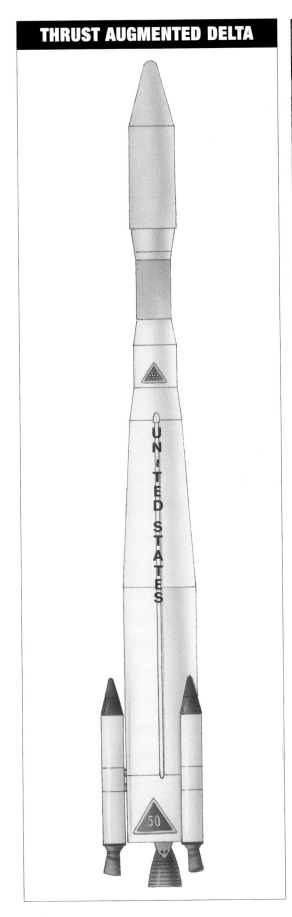

Soyuz – are all based on a first stage that is, in effect, an intercontinental or intermediate-range ballistic missile developed between 1957 and 1962. The major difference today is that whereas previously the capability of the launcher would dictate the size of a satellite, now the size of the satellite and the orbit into which it is to be delivered dictate the capability of the launcher. Launchers are now developed to serve a market, rather than the market being dependent on the launcher.

In the past, the satellite and launch industry used to be controlled by

Above: The core stage of the new Boeing Delta III booster is hauled into position on its launch pad No. 17 at Cape Canaveral. It is then mated to solid rocket strap-on boosters and its satellite payload.

Left: The Thrust Augmented Delta (TAD) had solid rocket boosters.

Above: A new liquid-propellant rocket motor is tested in California as part of a programme to develop new commercial, environmentally-friendly boosters.

Right: The Atlas Centaur was based on the Atlas ICBM, with a high-energy, liquid-oxygen, liquid-hydrogen upper stage.

governments and funded by taxpayers' money. Today much of the satellite and launch industry is commercialized. The transition from government control to a preponderance of private ownership has been a long one. The first Space Shuttle was launched in 1981 and the first NASA commercial deployment mission was STS 5 in November 1982, which sent two satellites into low Earth orbits. Further commercial deployments were made, and some stranded satellites were retrieved for refurbishment. It looked as though the Shuttle programme was well on its way. It was monopolising the launch business and many unmanned launchers were about to be scrapped. The turning point came in 1986 when the Space Shuttle Challenger exploded. The US had taken a risk in maintaining just one launch system. That decision backfired quickly when the orbiter Challenger was destroyed – and the lives of seven crew were lost. The Space Shuttle was grounded.

Without the presence of the Shuttle in the commercial sector, many US launchers were spared and a vibrant private launcher industry was born. Additional competition came, for example, from countries in Europe, even Russia. The collapse of the Soviet Union created a new economic

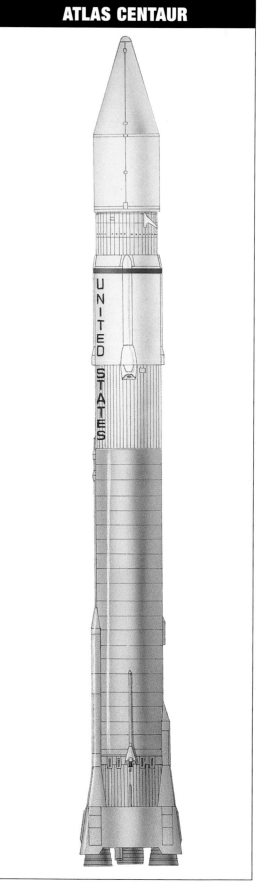

ATLAS CENTAUR

UNITED STATES

DELTA

The Thor Delta family grew in size as heavier satellites were launched, which prompted the development of more powerful solid rocket boosters.

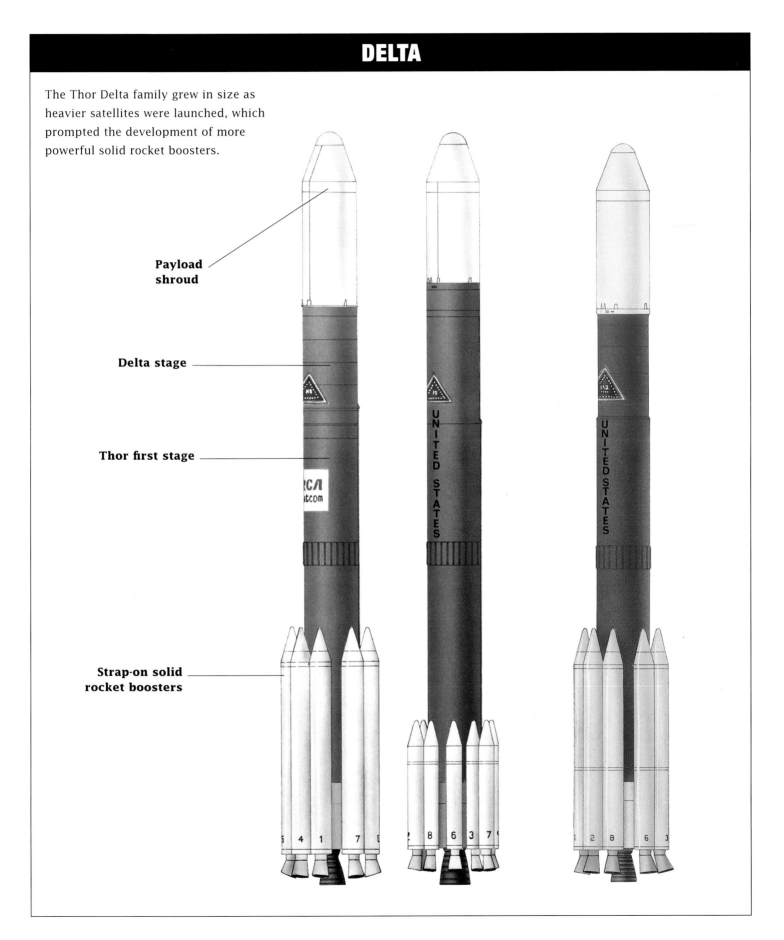

Payload shroud

Delta stage

Thor first stage

Strap-on solid rocket boosters

TITAN IIIB AND IIIC

A Titan II missile can be mated with an added Agena stage to produce the Titan IIIB Agena (left), and the Titan IIIC (right), which incorporates huge boosters.

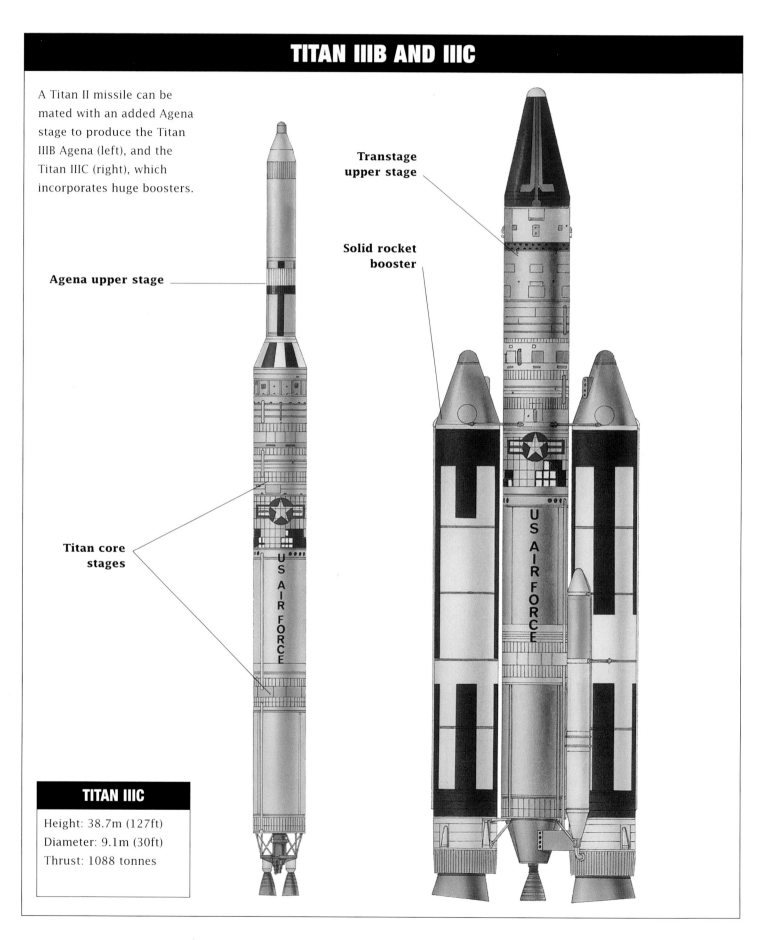

Transtage upper stage

Solid rocket booster

Agena upper stage

Titan core stages

TITAN IIIC

Height: 38.7m (127ft)
Diameter: 9.1m (30ft)
Thrust: 1088 tonnes

climate, and financial co-operation between the US and Russian space industries is such that the latest US launcher, Atlas V, will be powered by a Russian rocket engine! This represents a complete reversal of the early days of the Cold War when the Space Age was born.

MISSILES TO SPACE

The first steps towards modifying Cold War missiles, some of which placed the first satellites in orbit, was to equip them with upper stages to give them more capability. The Russian ICBM that launched Sputnik 1 was given an upper stage to send Luna probes to the Moon, and was gradually upgraded to enable it to launch satellites and manned spacecraft. It became known as the Vostok, Soyuz and Molniya in its various configurations. Today, the current Soyuz U model has flown almost 700 missions, including a record 100 consecutive flights without a failure. It has been equipped with two new stages, the Ikar and Fregat, which enable it to serve an international launcher market. It is marketed commercially by a joint Russian-French organisation called Starsem.

The US Atlas ICBM was equipped with a series of upper stages, including a veritable workhorse, the Agena. The Agena also flew as an upper stage for Thor and Titan missiles and was still being used in 1984. The Atlas was periodically upgraded with more powerful engines and also given a high-energy upper stage, called the Centaur, which flew successfully on a test flight in 1963. The Atlas Centaur is still flying today as an upgraded and improved model in a fleet of Atlas II boosters operated by the commercial organisation, International Launch Services (ILS). The Centaur is also being used on new Atlas models.

The Thor IRBM flew with a variety of upper stages, mainly the Agena and Delta. In 1963, another modification was introduced to improve the launcher's

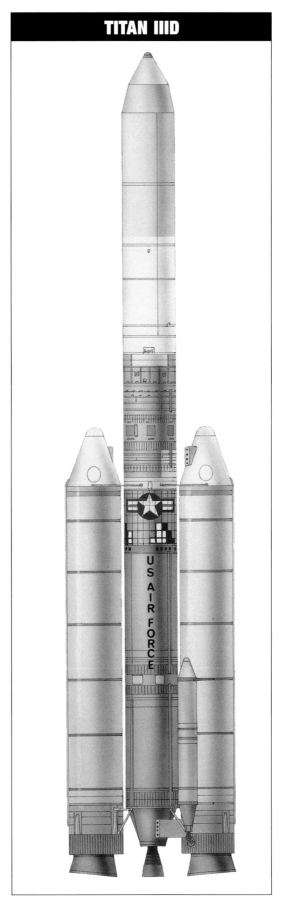

TITAN IIID

The collapse of the Soviet Union created a new economic climate, and financial cooperation between the US and Russian space industries is now such that the latest US launcher, Atlas V, will be powered by a Russian rocket engine.

Left: The Titan IIID incorporates a huge payload shroud to accommodate a reconnaissance satellite.

101

TITAN IIIE AND TITAN 34D

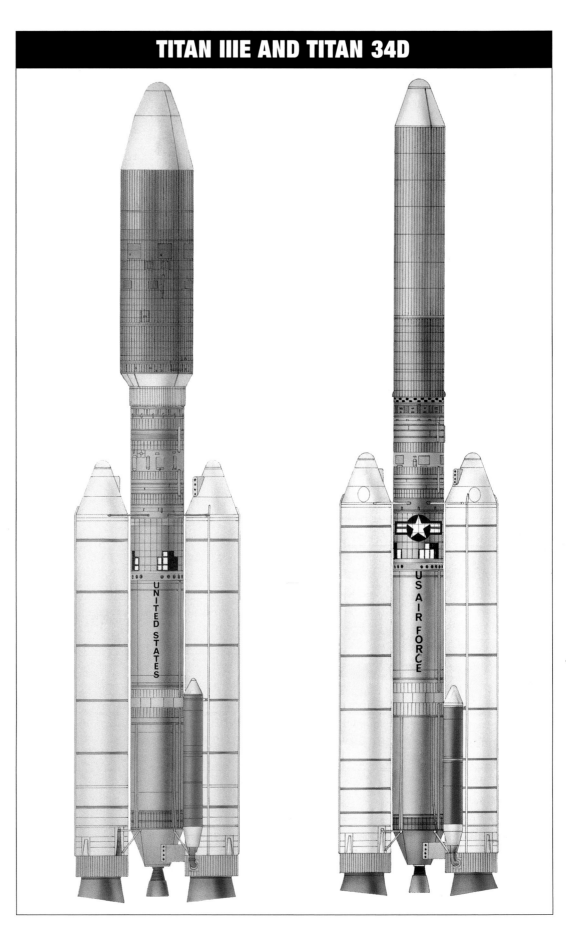

Right: The uprated Titan IIIE booster was developed for interplanetary launches, while the Titan 34D was designed to accommodate the heavier reconnaissance satellites.

performance. Instead of a more powerful upper stage, the Thor first stage was given an additional boost from solid-propellant, strap-on boosters. Today this is a successful means of providing a launcher fleet with a variety of launch capabilities, by using various combinations to produce different thrust levels. For example, today's Delta II – a direct descendant from Thor Delta – is available with a combination of nine, five or just three solid boosters, depending on the mission being flown.

The Titan II second generation ICBM was flown in 1962, and was soon in the satellite launcher business, as the core stage of a family of Titans with huge, solid-propellant strap-on boosters. It still flies today as the core stage of the Titan IVB-Centaur, the most powerful unmanned booster in use. With the introduction of Delta IV and Atlas V fleets by the US Air Force, the Titan will eventually be withdrawn from service. The Delta IV and Atlas V will also introduce the first "stacked first stage" approach, by flying three first stages together on some models. This method is also being adopted by Japan for its H2A launcher fleet to serve the commercial launcher market.

DELTA PERFORMANCE GROWTH

The Thor Delta launcher, which made its maiden flight in 1960, was the first of a family of Delta launchers which, with the new Delta IV fleet under development, will continue operations well into the new millennium. The Thor Delta was developed by the Douglas Aircraft Corporation which later merged with the McDonnell corporation to become McDonnell Douglas, which in turn was merged in the 1990s into the giant Boeing company of today. The Thor Delta was 30m (98ft) tall with the long slender payload fairing version. Its loaded lift-off weight was 114.17 tonnes. The Thor Delta was based on the Thor IRBM DM-21 first stage, 18.18m (59ft 8in) tall with a maximum diameter of 2.43m (8ft). The span over its tail fins was 3.9m (12ft 9in).

The 4.86 tonne first stage was powered by a single liquid oxygen-kerosene Rocketdyne MB3 engine with a thrust of 78 tonnes, and a burn time of two minutes 26 seconds. The second stage was 6.29m (20ft 8in) tall and 0.8m (32in) in diameter, weighing 2.69 tonnes, and was powered by an Aerojet General AJ10-118D engine using IRFNA UDMH hypergolic propellants, with a thrust of 3.43 tonnes and burn time of two minutes 50 seconds. The third stage was 1.52m (5ft) tall and 0.45m (18in) in diameter, and was powered by a solid propellant ABL X-248-A5 DM engine with a thrust of 1.25 tonnes and a 46 second burn time. The Thor Delta could place a payload weighing 226kg (498lb) into a 480km (298 mile) circular orbit. The Thor Delta soon became known just as the Delta which took various uprated forms and flew over 200 missions until the arrival of the Delta III in 1998. This booster has a poor launch record, with two failures in two attempts and will be superseded by the Delta IV in about 2002. The Delta III is 39m (128ft 2in) long and looks almost top-heavy with its enlarged second stage and payload shroud, 4m (13ft) in diameter. The booster weighs 301.45 tonnes at lift-off and can carry a payload weighing 3.81 tonnes into geostationary transfer orbit. The first stage is essentially an uprated Thor IRBM, powered with a Rocketdyne RS-27A engine with a thrust of 889,600N, burning liquid oxygen and kerosene propellants for 301s. The first stage thrust is augmented by nine huge solid rocket boosters 14.7m (48ft 2in) high which burn for four minutes 33 seconds. The second stage is powered by an engine derived from the Atlas Centaur booster. The RL-10B-2 Centaur cryogenic engine burns high-performance liquid oxygen and liquid hydrogen for 7 minutes 42 seconds, with a thrust of 110,094N.

NEW LAUNCHERS

The commercial market is led by Arianespace, a European consortium which

The Thor Delta launcher, which made its maiden flight in 1960, was the first of a family of Delta launchers which, with the new Delta IV fleet under development, will continue operating well into this millennium.

Above: The most powerful unmanned US booster, the Titan IVB Centaur, launches the NASA Cassini spacecraft towards Saturn from Pad No. 41 at Cape Canaveral.

operates a fleet of highly successful Ariane 4 and 5 rockets. The first Ariane flew in 1978 and today its successor, the Ariane 4, is offered in various combinations, with a mix of liquid and solid-propellant strap-on boosters, whilst the new and more powerful Ariane 5 has two large solid-propellant rocket boosters, known as SRBs. Through the Ariane fleet, Arianespace can offer a range of seven boosters, and a payload ranging from 2.1 to 6.8 tonnes to geostationary transfer orbit. With this flexible capability, Arianespace's launch order book usually numbers about 40 satellites at any one time.

ARIANE I AND ARIANE 3

Launched on 24 December 1979, Ariane 1 was the first in a series of highly successful European commercial launches. The first launch contained a technological capsule to instead of a satellite to monitor the rocket's behaviour. Ariane 3, first launched in 1984, had improved thrust in all three stages.

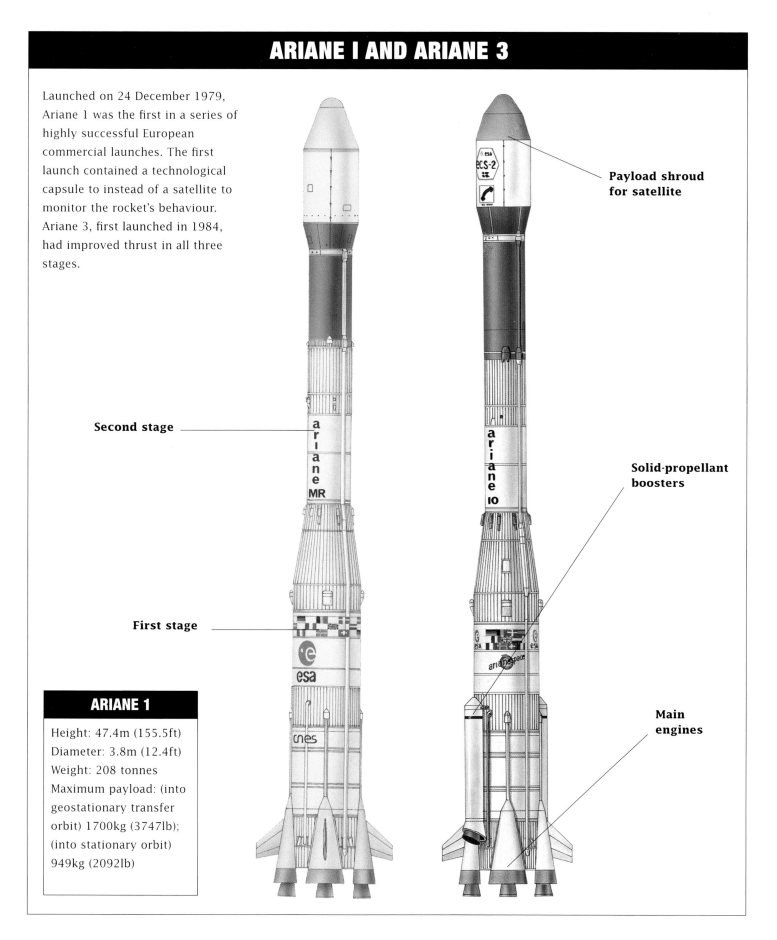

Payload shroud for satellite

Second stage

Solid-propellant boosters

First stage

Main engines

ARIANE 1

Height: 47.4m (155.5ft)
Diameter: 3.8m (12.4ft)
Weight: 208 tonnes
Maximum payload: (into geostationary transfer orbit) 1700kg (3747lb); (into stationary orbit) 949kg (2092lb)

ARIANE 4 AND ARIANE 5

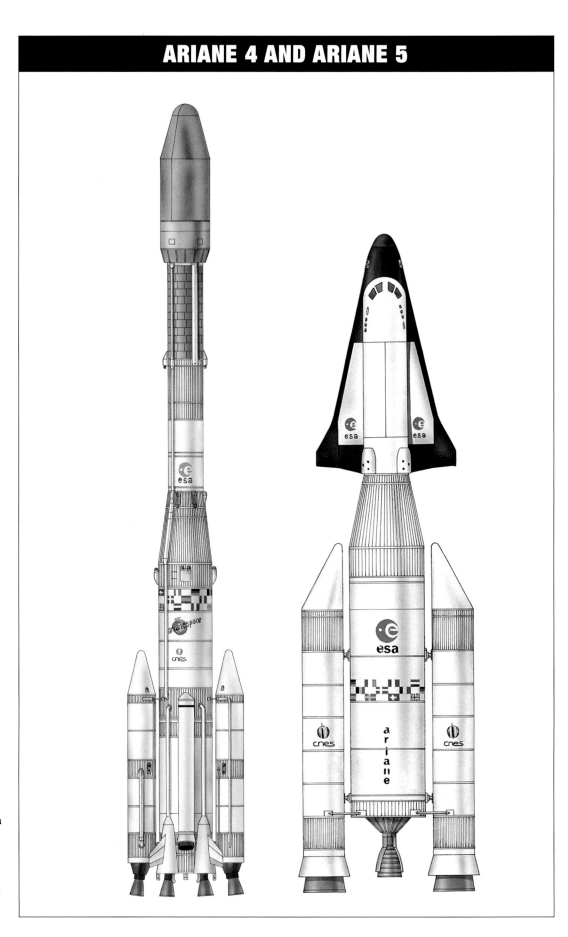

Right: The Ariane 4 was first launched in 1988, while the Ariane 5, whose first launch took place in 1997, was originally developed to launch the Hermes mini-shuttle as shown here. Hermes was subsequently cancelled, and Ariane 5 has been adapted for other purposes.

Russia is a partner in the International Launch Services, offering its Proton booster to complement the American Atlas. The Proton, which can place 2.6 tonnes directly in geostationary orbit using a restartable upper stage, first flew in 1967 and has clocked up almost 250 launches. A new Russian vehicle called Anagara is being introduced to eventually replace the Proton. Its design will include the "stacked" approach of linking core stages together to offer a flexible payload range to customers at home and abroad. Russia also operates a fleet of other launchers, including the Zenit 2, using the technology of the strap-on boosters of the Energia heavylift-booster

Left: Ariane 4 was the workhorse of the Arianespace fleet, and was available in six configurations. The 42L had two liquid-propellant strap-on boosters.

HOW ROCKETS WORK

Watching a rocket blast-off, it is easy to imagine that it is the thrust of the rocket engine pushing against the ground, then the air beneath it, that propels it upwards. But that is not so. If that were the case, how could a rocket will also work in the vacuum of space where there is no airlift? Blow up a balloon, let it go, and you will see it rush through the air. It is exhaust that propels a rocket. As the fuel burns, the exhaust is forced out of a carefully designed nozzle which is controlled by such guidance systems as gyroscopes. Three centuries ago, Isaac Newton explained "for every reaction, there is an equal and opposite reaction." Action-and-reaction is the principle of the rocket engine. In fact, a rocket engine works rather like a "controlled explosion." The most common space-rocket engine burns fuel, such as kerosene, with an oxidizer, liquid oxygen, in a combustion chamber. This activity creates very hot gases which are under enormous pressure. The pressurized gases are given an escape route – a small, restricted exit at the back of the combustion chamber. Forcing the pressurized gases through a small exit accelerates the gases, providing extra thrust.

A conical nozzle, fixed to the throat of the exit of the combustion chamber, causes the gases to accelerate even more because their flow is restricted still further. The controlled direction of the exhaust from the nozzle gives the rocket some flight control.

Providing a safe and efficient way to mix and burn the fuel and oxidizers requires careful design. In the Space Shuttle's main engine systems, for example, the oxidizer and fuel – in this case liquid hydrogen rather than kerosene – are first pressurized, mixed and pre-burned to form hot gases. Then, in a precise mixture, the gases are introduced into the combustion chamber. Liquid oxygen and hydrogen engines, called cryogenic engines, are more powerful than traditional engines, such as the American Atlas, which use liquid oxygen and kerosene. Some other liquid-propellant rockets use different chemicals. Nitrogen tetroxide oxidizer and hydrazine fuel, when mixed, ignite spontaneously without needing an ignitor. They are called hypergolic engines. The core first stages of the US's Titan boosters use these engines. All burning of liquid propellants can be controlled carefully in flight and, if necessary, the engines can be shut down. That is not the case with solid-propellant motors which are commonly used on military missiles, and used widely in the space industry to provide extra thrust during the early phases of a rocket launch. The Shuttle, for example, uses two mighty candle-like SRBs which are jettisoned after two minutes. If anything goes wrong they cannot be turned off. The American Delta uses up to nine smaller strap-on boosters, mounted around the base of the thrust stage. Hypergolic engines and solid-propellant motors are commonly used in the upper stages of rockets. They are also used as the propulsion systems on spacecraft proper, to adjust their orbits and control altitude.

Right: The Russian RD-180 engine has been developed with the USA to power the first stage of the new Atlas III booster, which made its first flight in 2000.

PROTON D-1

The Proton D-1 was launched in 1965 and was the first in a series of boosters which launch commercial communications satellites.

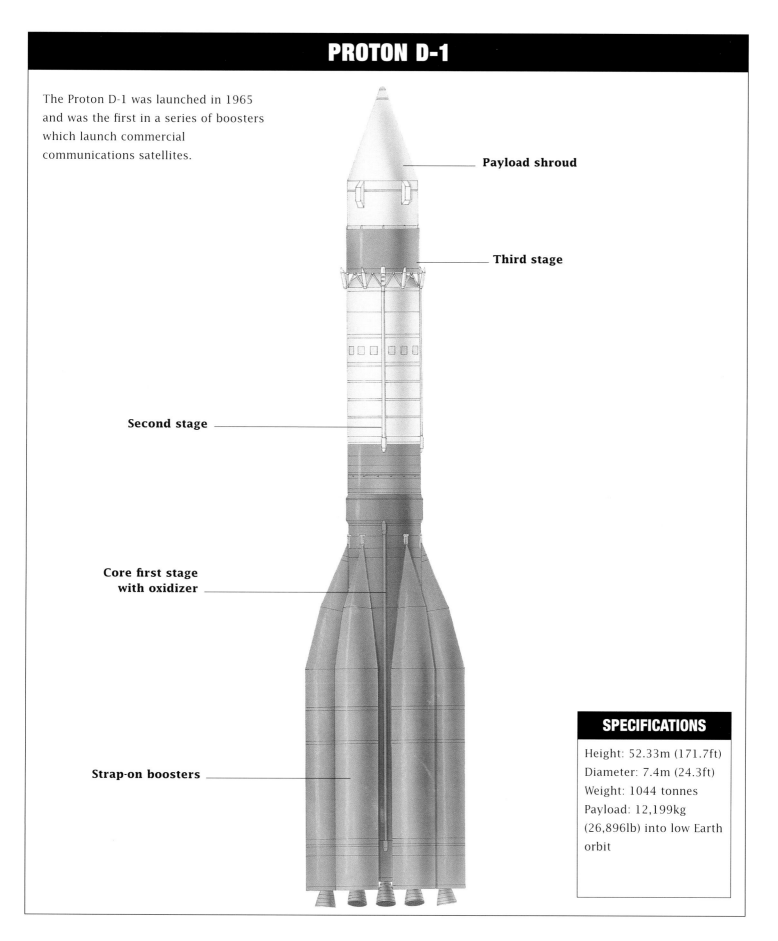

Payload shroud

Third stage

Second stage

Core first stage with oxidizer

Strap-on boosters

SPECIFICATIONS

Height: 52.33m (171.7ft)
Diameter: 7.4m (24.3ft)
Weight: 1044 tonnes
Payload: 12,199kg
(26,896lb) into low Earth
orbit

COSMOS B-1 AND C-1

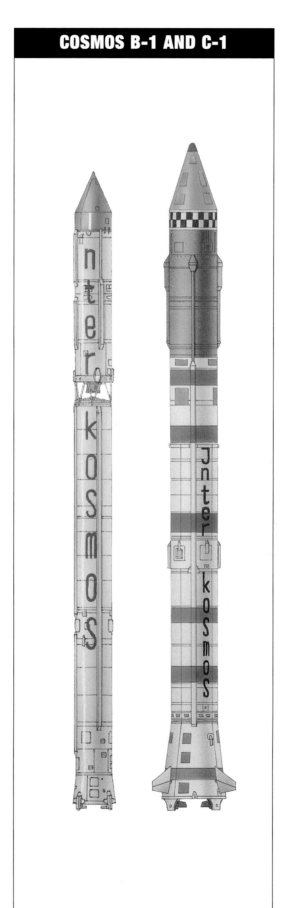

Far right: The workhorses of the International Launch Services company (ILS) are the Proton and the Atlas. The Proton is seen launching a communications satellite for Inmarsat mobile communications company.

Right: The Cosmos B-1 was first launched in 1962, followed by the C-1 two years later. They are based on the designs for the Sandal and Skean missiles.

developed for Russia's cancelled space shuttle programme. Other launchers include the Tsyklon, based on the SS-9 Scarp ICBM, which first flew in 1996, and the Cosmos, based on the SS-5 Skean IRBM. This first flew in 1964m, following an original Cosmos version of 1962 based on the SS-5 Sandal IRBM. Another Russian former missile is the SS-19 which has been converted into the Rokot satellite launcher

marketed by a Russian-German company called Eurokot. As part of the SALT 2 treaty with the US, Russia has scrapped hundreds of missiles and is hoping to convert many others into satellite launchers.

OTHER NATIONS IN SPACE

Another participant in the commercial launcher market is China, which launched its first satellite in 1970 using a modified ICBM called the Long March 1. Today, China offers a range of Long March boosters for national and customer use. The Long March 3B can launch payloads weighing 5 tonnes to geostationary transfer orbit. China offers its launchers at a budget price of about $40 million compared with the $80 million to $100 million charged by western launchers. However, the number of its customers is limited by international controls, and a reputation for some spectacular failures, some of which have resulted in the loss of western customers' satellites.

Satellite operators have to pay more than the launch price to get into orbit. A typical satellite will cost about $150

Above: These stages of a Russian Proton commercial booster are being prepared for the construction of a new vehicle for launch. The Proton is operated by a joint US-Russian company.

Leftt: The China Great Wall Industry Corporation operates a series of Long March boosters, including this Long March 2E seen here being launched from Xichang.

THE LONG MARCH 1 AND 2C

The Long March 1, based on China's ICBM, launched China's first ever artificial satellite in on 24 April 1970. First launched five years later, the Long March 2C is still in use today. It is built at the Shanghai Xinxin Machine Factory, where further launchers are being developed in the Long March series as China strives for an international market. China has tried to overcome the lacklustre reputation of its launchers in the West by offering them at extremely competitive prices.

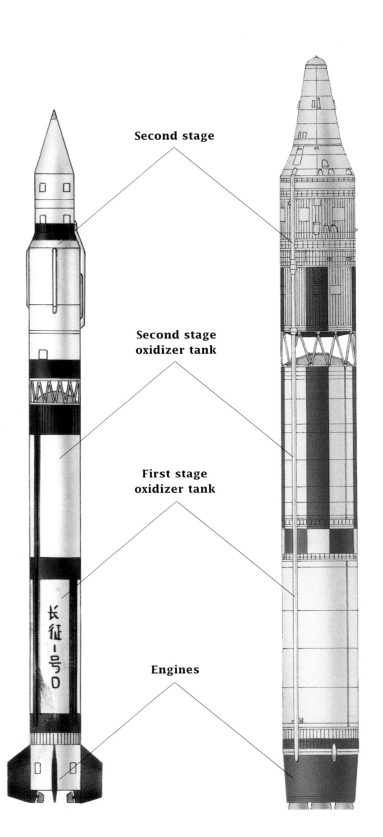

Second stage

Second stage oxidizer tank

First stage oxidizer tank

Engines

LONG MARCH 2C

Height: 35m (115ft)
Diameter: 3.35m (11ft)
Weight:191 tonnes
Payload: 2199kg (4850lb) into low Earth orbit
Range: up to 5000km (3106 miles)

Left: A Japanese H2 booster is launched from Tanegashima, carrying a commercial communications satellite into geostationary orbit.

million and, added to the launch price, the bill can reach $250 million. To this must be added the important launch insurance premium which can push the bill up to $350 million or more. The satellite-launch insurance business is a major part of the industry, and premiums will depend on the performance of the launchers and the

JAPAN

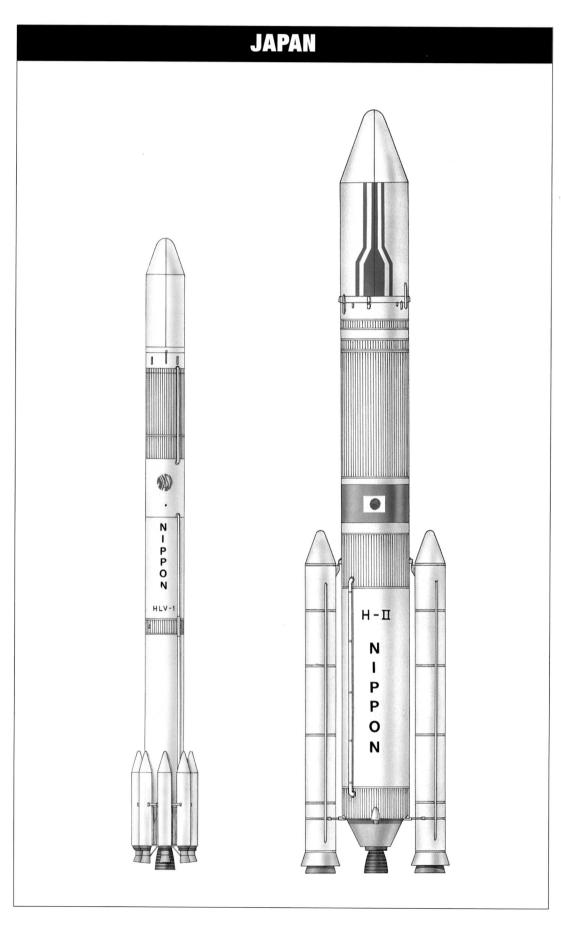

Right: Japan's H1 was designed to place 550kg (1212lb) into geostationary orbit, while the powerful H2 was intended to challenge not only the Ariane 5, but also the best US commercial launchers.

SATELLITE LAUNCHER SITES

Date	Launch site
4 October 1957	Tyuratam/Baikonur, Kazakhstan, Soviet Union
31 January 1958	Cape Canaveral, Florida, US
28 February 1958	Vandenberg, California, US
16 February 1961	Wallops Island, Virginia, US
16 March 1962	Kapustin Yar, Russia, Soviet Union
26 November 1965	Hammaguir, Sahara (France)
17 March 1966	Plesetsk, Russia, Soviet Union
26 April 1967	Italian San Marco mobile platform (US launch)
9 November 1967	Kennedy Space Centre, Florida, US
29 November 1967	Woomera, South Australia (US launch)
11 February 1970	Kagoshima, Japan
10 March 1970	Kourou, Guiana, South America (French launch)
24 April 1970	Jiuquan, China
9 September 1975	Tanegashima, Japan
18 July 1980	Sriharikota, India
29 Jan 1984	Xichang, China
19 September 1988	Palmachin AFB, Yavne, Negev, Israel
5 April 1990	B52 air launch over Pacific, US
3 September 1990	Taiyuan, China
3 April 1995	TriStar L-1011 air launch over Atlantic Ocean
4 March 1997	Svododny, Russia
7 July 1998	Submarine launch, Barants Sea, Russia
28 March 1999	Sea Launch Odyssey Platform Pacific

Above: The Chinese Long March 4 is launched from Taiyuan carrying a national weather satellite. Some Long March boosters are being marketed commercially.

Left: The Sea Launch Zenit 3SL booster is prepared for tests before being shipped on its offshore platform to the mid-Pacific to launch a satellite.

Right: India is developing a geostationary satellite launcher based on some of the components of this Polar Satellite Launch Vehicle at Sriharikota.

SPACE DEBRIS

The sad fact is that spent rocket stages abandoned in orbit represent the biggest danger to using space in the future. Space debris is a perennial problem that will not go away. About 8700 man-made objects larger than tennis balls – of which only about 700 are operational satellites – can be tracked in Earth orbit. Of these objects, 41 per cent are large fragments left over from explosions, usually caused by the ignition of unspent propellant in rocket stages. Discarded upper stages represent 17 per cent of debris; defunct satellites, 22 per cent; and ejected payload shrouds and covers, together with tools left over by spacewalking astronauts, and other items, account for 13 per cent. Their number is increasing all the time while debris smaller than the size of a tennis ball, as little as 1cm, is estimated to number over 150,000. A 1cm size fragment orbiting at a speed of more than 28,000km/h (17,400mph), or 8km (5 miles) per second, could shatter a $100 million satellite – or the Space Shuttle.

Spacecraft are also being peppered by some of the 34,000 micron-sized particles (each one a thousandth of a millimetre) that exist in space. Many of these particles of debris are made of aluminium oxide issued from the solid-propellant rocket motors on launchers. A tiny fragment could easily puncture an EVA spacesuit. A window of the Space Shuttle Challenger was chipped by a 0.3mm fleck of paint which impacted at four metres per second. In the first known collision of space objects, in 1996, the French Cerise satellite's long antenna was severed by a fragment of an Ariane 4 third stage, hitting at a speed of 14km per second (8.7 miles per second. The debris problem can be reduced. One of several ways of preventing the build-up of further debris would be to vent into space any residual fuel left in the tanks of some stages. This reduces the risk of the rocket stage exploding, thereby creating more debris. Arianespace has already introduced this as an operational procedure during launches. Stages could also deliberately be de-orbited and brought safely back to Earth after completing their work.

It is likely that other countries such as North Korea and its neighbour to the south will be joining the space "club" within the next few years, as satellite technology proliferates around the globe.

record of the industry in terms of launch success. In 1999, several launch failures cost the market $1 billion in payouts, which exceeded premiums received. Such a situation will inevitably push the insurance rates up even further.

Other countries, including Japan, Brazil, India and Israel, operate satellite launchers. Brazil has not yet launched a satellite using its VLS booster which has failed on two launch attempts. The VLS is a small launcher, capable of placing only 200kg (441lb) in low Earth orbit, so it is unlikely to be offered commercially. Japan launched its first satellite in 1971 and now operates the H2, which is being offered as a commercial launcher for communications satellites.

India operates a Polar Satellite Launch Vehicle which has made one commercial flight, carrying piggyback satellites with a national remote-sensing satellite payload. India is also developing its own Geostationary Satellite Launch Vehicle. Israel operates a Shavit booster for low earth-orbit launches, which may be combined with US technology to form commercial boosters. The Shavit is based on Israel's military missile. North Korea is reported to be planning to launch a satellite using a version of its Tapeo Dong 2 missile. It is likely that this country and others, including its neighbour South Korea, will be joining the space "club" within the next few years, as satellite technology proliferates around the globe.

THE SPACE SHUTTLE

Arguably the most famous space vehicle today, the Space Shuttle has flown almost 100 missions since 1981. Although the Shuttle has not flown as frequently as planned when the programme began in 1972, it has made space travel appear almost routine. There is simply no other space transporter like it.

The Space Shuttle comprises three main elements: the orbiter which is equipped with three massive engines called the Space Shuttle main engines (SSMEs), which are fuelled by liquid oxygen and liquid hydrogen fed from a large brown external tank attached to the belly of the orbiter. The external tank (ET), which was painted white for the first two missions in 1981, and is now the familiar brown colour, is the only part of the Shuttle that is expendable, being abandoned to re-enter Earth's atmosphere when the orbiter has reached initial orbit. Attached to either side of the external tank are two solid rocket boosters (SRBs) which supplement the SSMEs during the first two minutes of flight, and

Left: View of a Shuttle launch taken by an automatic camera on the beach near the launch pad. The Shuttle orbiter is obscured by the external tank and solid rocket boosters.

which are recovered from the sea after each flight. Most component parts of the SRBs are used again on future Shuttle flights.

PAYLOAD CAPABILITIES

The Shuttle's payloads are carried primarily in the payload bay which measures 18.3m (60ft) long and 4.6m (15ft) wide. Additional cargo and experiments, as well as crew equipment and consumables, are located in the mid-deck which is situated under the flight deck. The mid-deck also acts as the wardroom, kitchen and temporary gym, and is equipped with a toilet. Astronauts can sleep in bunks in the mid-deck but many choose their own places, such as in

the airlock – normally used as an entrance to attached Spacelab and Spacehab modules, and for EVAs – or in the flight deck. Spacelab is a laboratory which is fitted inside the payload bay, and which has been used many times for various science missions. Spacehab provides additional working, storage and instrument space for some Shuttle missions. Like Spacelab, it extends into the payload bay. For Shuttle missions which dock to space stations, the airlock leads to a docking module.

The Shuttle's original maximum capability for carrying a payload to a low earth-orbit of 28.5˚ inclination was

Above: The Shuttle Endeavour and its mobile launch platform reach launch Complex 39, where it will be enveloped by the rotating service structure on the left while being prepared for launch.

Far left: A Space Shuttle is rolled the 4.83km (3 miles) to the launch site on a mobile launch pad after its assembly in the Vehicle Assembly Building at the Kennedy Space Centre.

121

THE BIRTH OF THE SHUTTLE

Above: Many versions of the Space Shuttle were proposed before the vehicle was built, including this version based on the Saturn V first stage.

The concept of a re-usable space "plane" was entirely logical. Wernher von Braun and his Peenemünde rocket team were planning to develop a winged, recoverable "space shuttle" vehicle, based on V2 technology. Many artists' early concepts of future space travel featured winged, aeroplane-like rocketships. The US flew re-usable manned rocket planes in the 1950s and 1960s, including the legendary X-15 which reached the edge of space. If there had not been so much urgency, influenced by Cold War politics, perhaps a space shuttle vehicle would have been developed sooner. The nearest the US came to it was a concept called Dyna Soar, a winged, glider-like space plane which would be launched aboard a Titan 3C booster for manned military missions for the US Air Force. Six astronauts had been selected to fly it before the project was cancelled. The priority of the Space Race was to get into space first and as a result, manned capsules, capable of making just one flight, became the name of the game.

The idea of developing a true space-shuttle-type vehicle was not seriously addressed by NASA or politicians until Apollo 11 had reached the goal of landing on the moon. There were to have been nine further Apollo missions, but in the anti-climax and inevitable reduction in budgets, three missions were cancelled. In early 1970, the next goal of an ambitious NASA was to build a space station and use a "taxi" to ferry cargo and crew to and

from it, and make space travel as routine as flying. The White House and Congress were not so enthusiastic, however, as the budget for these projects would be more than Apollo's entire budget of $25 billion. NASA was forced to cancel the huge space-station project and was left with a "space taxi" with nowhere to go.

The "taxi" was renamed the Space Shuttle by NASA and advertised as a fully re-usable, versatile system that could carry commercial satellites into orbit, charging a fare; could act as a mini-space station and laboratory; carry out space repairs; and accomplish many more space tasks. However, in doing this, NASA was compromising the Shuttle's competitiveness as a launcher, and its capability as a spacecraft. NASA also claimed that the Shuttle would fly 20 times before 1980. An extraordinary 650 missions would have been flown by 1991.

Many aerospace firms had developed designs for the new vehicle. One of the most attractive was a two-stage, winged-booster space plane which could be flown both manned and unmanned depending on the mission; the booster would be equipped with air-breathing engines for operating in the atmosphere. There was a problem, however. The budget for the Space Shuttle was going to be extremely limited – the equivalent of about one-fifth of the cost of Apollo, in order to build a system that would be five times more challenging than Apollo. The result was predictable. President Richard Nixon gave the Space Shuttle the go-ahead in January 1972. By July of the same year the former Rockwell company (now part of Boeing) was contracted by NASA to build the Shuttle. The chosen design was not fully re-usable as a matter of financial necessity, and effectively an engineering compromise.

THE SHUTTLE ORBITER

Six Shuttle orbiters were built. Enterprise was used for testing only and did not fly in space. Columbia was followed by Challenger, Discovery and Atlantis, and Endeavour was built to replace Challenger after its tragic loss.

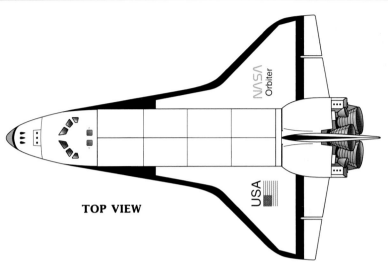

TOP VIEW

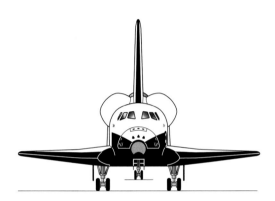

FRONT VIEW

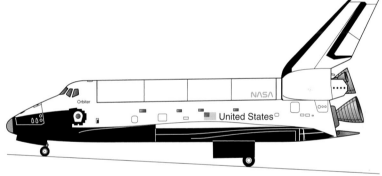

SIDE VIEW

REAR VIEW

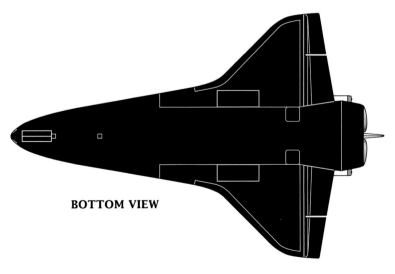

BOTTOM VIEW

MINIMUM GROUND CLEARANCES

Body flap (aft end): 3.68m (12.07ft)

Main gear (door): 0.87m (2.85ft)

Nose gear (door): 0.90m (2.95ft)

Wingtip: 3.63m (11.92ft)

advertised as 29.48 tonnes, but the heaviest payload to be carried to date weighed 23.72 tonnes, and this was on the ill-fated Challenger which was lost following its launch in 1986. The heaviest payload since then is the 22.58 tonnes on STS 93 Columbia in 1999. The updated maximum payload capability is now 24.95 tonnes, and that is for a basic four-day, satellite deployment mission in minimum orbit. However, this payload weight has not yet been carried on a Shuttle. After the Challenger accident exposed not just poorly designed SRBs but launch-dynamic loads far in excess of those predicted when the Shuttle was first designed – a fact which was extremely well-hidden by NASA – the whole system for launch was redesigned and strengthened. The weight of the total Shuttle system was increased, and resulted in a reduction in payload capability.

The payload capability depends on the orbital inclination taken by the mission. Each degree higher than 28° takes 226kg (498lb) off the payload weight. Missions from Kennedy Space Centre can fly up the eastern seaboard of the US into a 57° orbit. On one reconnaissance satellite-deployment mission this angle was extended to 62°. Early in the programme, launches into polar orbits of 90° inclination were also planned from Vandenberg AFB in California. Just prior to the Challenger accident of 1986, a crew of seven was due to undertake a military mission aboard a Discovery, but the launch was cancelled. For safety reasons, all Vandenberg flights were subsequently cancelled.

Below: The Space Shuttle Enterprise was used in 1977 for a series of approach and landing tests (ALTs), in which the orbiter and its two pilots separated from a Boeing 747 carrier aircraft to glide onto a runway.

The original plan for the Shuttle was to be able to fly for 30 days. This has never been achieved, although a record 17-day mission was made possible by the use of a relatively new unit called the Extended Duration Orbiter (EDO). The EDO adds more fuel cells to the orbiter, thereby generating additional electric power. Other missions of a similar duration have also been flown with the EDO. Another early plan was to relaunch an orbiter within 14 days of completing a mission, but this turnaround has not been possible, mainly because processing takes much longer than planned, and more components need to be changed than was at first anticipated.

Six Shuttle orbiters have been built, beginning with the Enterprise in 1977. The Enterprise was not a spaceworthy craft and was used for atmospheric glide tests before the space missions could be attempted. Five Approach and Landing Tests (ALTs) were flown by two teams of two astronauts – Fred Haise and Gordon Fullerton, and Joe Engle and Dick Truly. The Enterprise flew mounted on a Boeing 747, and was then released to fly to a landing at Edwards Air Force Base in California. The longest of these flights, which took place between August and October 1977, lasted for more than five minutes. The first orbiter to make an orbital space flight was Columbia, in 1981; the second was Challenger, in 1983; the third, Discovery, in 1984; and then Atlantis in 1985. After Challenger was lost in 1986, a replacement orbiter, Endeavour, was built and flew for the first time in 1992.

UNIQUE ROCKET BOOSTERS

The orbiters are 37.24m (122.2ft) long, with a wing span of 23.79m (78ft) and a height of 17.27m (56.67ft), from the undercarriage to the top of the vertical stabilizer. When the Space Shuttle is on the launch pad and fully loaded for lift-off, it weighs around 2041 tonnes, of which the orbiter accounts for about 113 tonnes (5.54 per cent). The SRBs are 45.46m (149.15ft)

and 3.7m (12.14ft) in diameter. The external tank is 47m (154.2ft) long and 8.38m (27.5ft) wide. The total length of the whole stack, from the tip of the tank to the tail of the SRBs, is 56.14m (184.2ft).

The SRBs are the largest solid-propellant motors ever flown and the first designed to be re-used. They each weigh about 57 tonnes at time of launch, 85 per cent of that being the propellant, which comprises ammonium perchlorate oxidizer, aluminium fuel, ion oxide catalyst, a polymer binder and an epoxy-curing agent. The SRBs each develop a thrust of about 1497 tonnes and only ignite after the SSMEs

Above: The two solid rocket boosters are separated from the Space Shuttle two minutes after the launch of STS 2 in 1981.

SPACE SHUTTLE

The Space Shuttle comprises an orbiter, external propellant tank and two solid rocket boosters. Its capability varies from mission to mission.

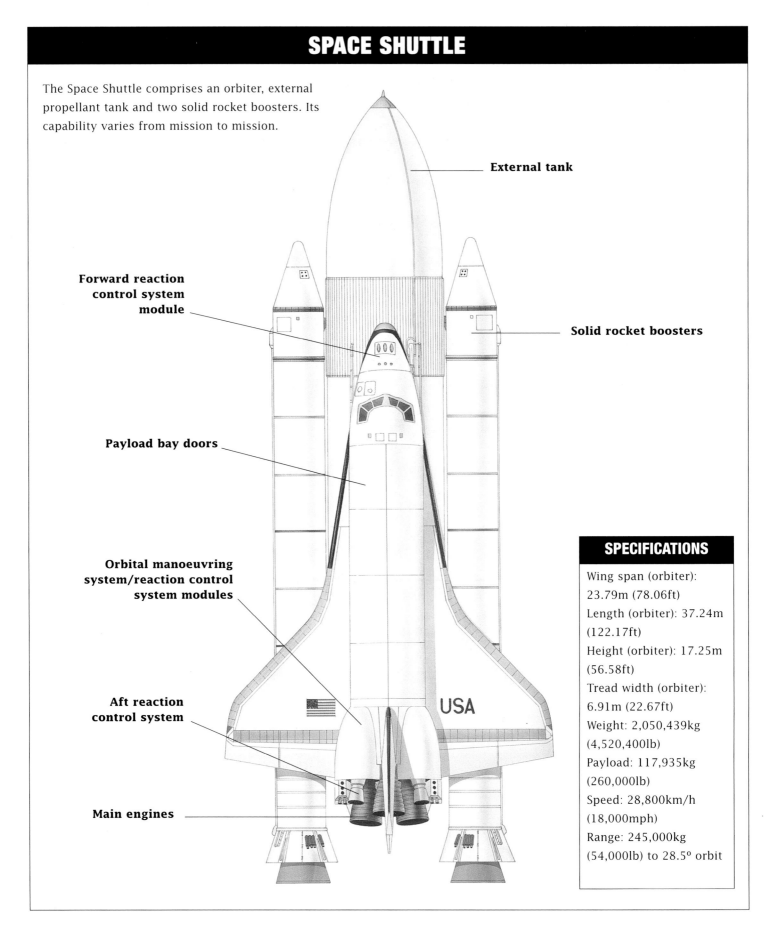

External tank

Forward reaction control system module

Solid rocket boosters

Payload bay doors

Orbital manoeuvring system/reaction control system modules

Aft reaction control system

Main engines

USA

SPECIFICATIONS

Wing span (orbiter): 23.79m (78.06ft)
Length (orbiter): 37.24m (122.17ft)
Height (orbiter): 17.25m (56.58ft)
Tread width (orbiter): 6.91m (22.67ft)
Weight: 2,050,439kg (4,520,400lb)
Payload: 117,935kg (260,000lb)
Speed: 28,800km/h (18,000mph)
Range: 245,000kg (54,000lb) to 28.5° orbit

have built up to full thrust to provide 71 per cent of the vehicle's thrust at lift-off. Like the SSMEs, the SRBs are throttled back – by a third, at T+50 seconds – to avoid over-stressing the vehicle during the period of maximum dynamic pressure, known as Max Q. A thrust-vector control system also enables the SRB nozzles to be steered. The SRBs burn for two minutes, and are then jettisoned at an altitude of about 44km (27.34 miles) with the aid of 16 separation motors. With the momentum of their velocity, the boosters coast upwards for about 75 seconds to an altitude of about 65km (40.4 miles) before falling back towards Earth. About 225km (140 miles) downrange, and at around four minutes 41 seconds after launch, they splash down in the Atlantic Ocean under three parachutes. The boosters are recovered, their five segments dissembled and returned to Utah for refurbishment and assembly.

SHUTTLE PROPULSION

Fully loaded, the heaviest element of the Space Shuttle is the ET which weighs 751 tonnes, of which 617 tonnes is liquid oxygen and 103 tonnes is liquid hydrogen. The weight of the ET was reduced twice after the launch of STS 1 in 1981. More recently, a further reduction by 3.5 tonnes to 747 tonnes was made possible by installing a new lightweight tank, which uses an aluminium-lithium alloy instead of aluminium alone; a lighter foam was also used to insulate the tank. Any reduction in the weight of the Shuttle system can be translated into more payload. The ET remains attached to the Space Shuttle until the SSMEs have cut off at an altitude of

Left: Soon after SSME shutdown, the external tank is released into the upper atmosphere to be destroyed.

MAIN ENGINES, SOLID ROCKET BOOSTERS AND EXTERNAL TANK

The three main engines are located in the aft section of the orbiter.They are supplied with liquid oxygen and liquid hydrogen propellants from the external tank. Each engine combines high chamber pressure operation, a bell-shaped nozzle and a regeneratively cooled thrust chamber for maximum performance. They are gimballed for thrust vector control while each engine is running. Each SSME has an engine controller based on a digital computer which monitors engine performance and automatically adjusts its operation for the required thrust and constant mixture ratio. Should one of the three SSMEs suffer a flameout, fuel would be diverted to run the remaining engines for a longer time. Spray-on foam insulation is applied over portions of the external tank to reduce ice or frost formation during launch, and to minimize heat leaks into the tank which cause boiling of the liquid propellants. The two solid rocket boosters, recoverable by parachute, fire in parallel with the SSMEs of the orbiter to lift the Shuttle from the pad into vertical flight.

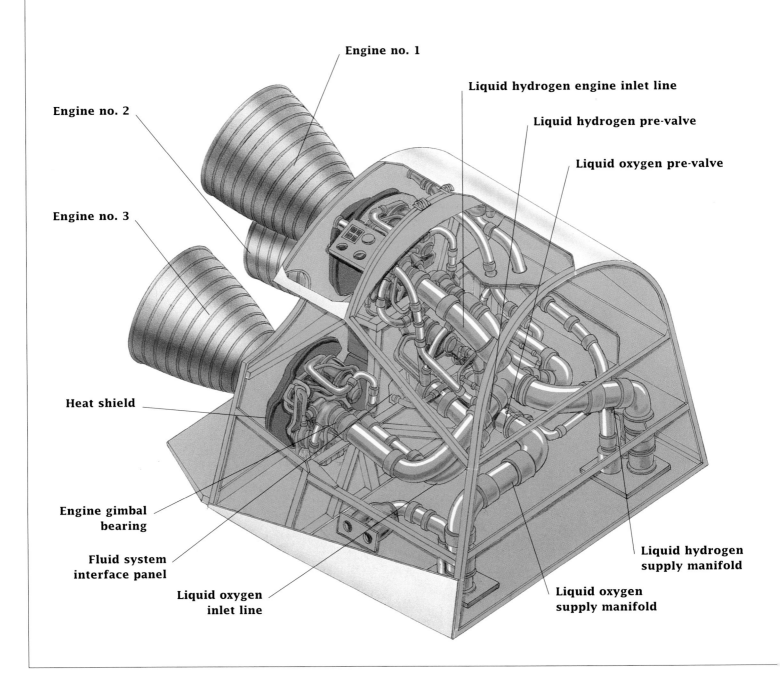

Engine no. 1

Liquid hydrogen engine inlet line

Liquid hydrogen pre-valve

Liquid oxygen pre-valve

Engine no. 2

Engine no. 3

Heat shield

Engine gimbal bearing

Fluid system interface panel

Liquid oxygen inlet line

Liquid oxygen supply manifold

Liquid hydrogen supply manifold

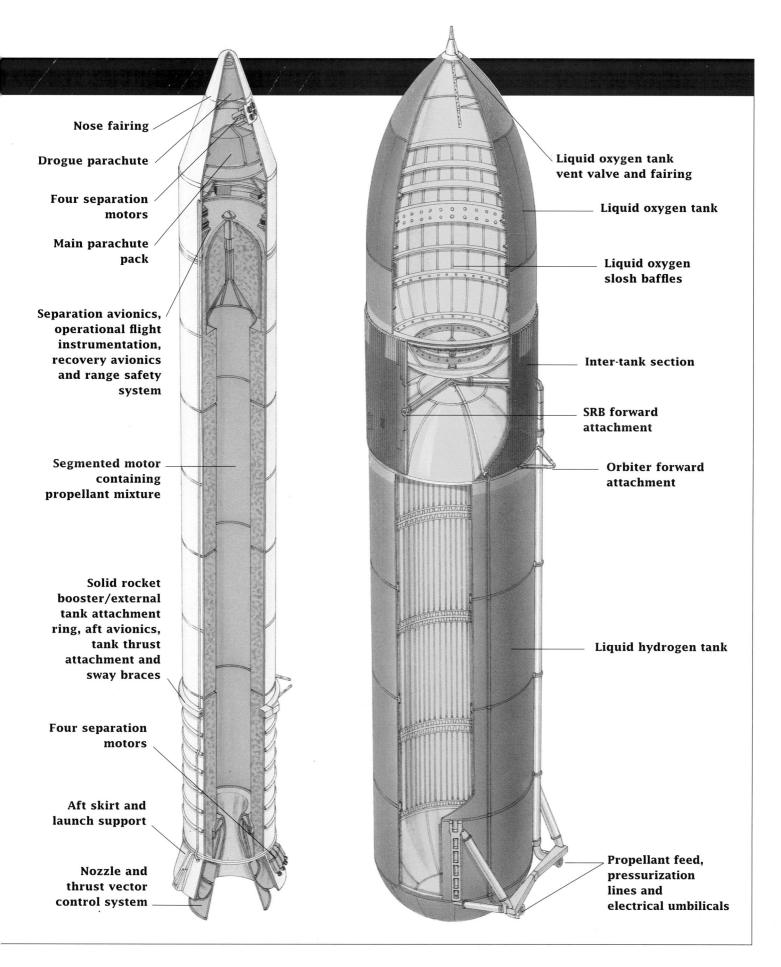

Nose fairing

Drogue parachute

Four separation motors

Main parachute pack

Separation avionics, operational flight instrumentation, recovery avionics and range safety system

Segmented motor containing propellant mixture

Solid rocket booster/external tank attachment ring, aft avionics, tank thrust attachment and sway braces

Four separation motors

Aft skirt and launch support

Nozzle and thrust vector control system

Liquid oxygen tank vent valve and fairing

Liquid oxygen tank

Liquid oxygen slosh baffles

Inter-tank section

SRB forward attachment

Orbiter forward attachment

Liquid hydrogen tank

Propellant feed, pressurization lines and electrical umbilicals

about 80km (49.7 miles), eight minutes after launch, when the ET is jettisoned. It then burns up in the Earth's atmosphere.

The orbiter's liquid oxygen-liquid hydrogen (LOX-LH) cryogenic SSMEs have been described as the most efficient rocket engine ever designed. This is borne out by the fact that only one engine has had to be shut down during a launch – STS 51F in August 1985 – and even that was put down to a faulty temperature reading. The propellants are "pre-burned" before being fed into the final combustion chamber, which increases temperature and pressure, leading to greater efficiency. The LOX-LH mix ratio of 6:1 is maintained by a computer with two back-ups. The SSMEs ignite in sequence, starting with No 3 at T-6.6 seconds. All three then reach full thrust

by T-3 seconds. If any engine has failed to reach this thrust by T-2 seconds, the engines are shut down automatically. This shutdown has occurred five times on launch attempts for a range of technical reasons which have prevented one of the engines from starting or being cut off. Launch pad aborts are dangerous since fire is a real possibility. If this happens, the crew would have to make an emergency egress by sliding down wires to an underground bunker or to an armoured truck for a rapid escape. Although evacuating for fire is a well-practiced procedure, no crew has yet had to do it for real.

The SSME exhaust nozzle is 4.26m (13.98ft) high and 2.43m (7.97ft) in diameter at the exit. The SSMEs can be

Above: The Space Shuttle has reached orbit and the suited commander, Robert Gibson, can relax a little before the mission starts in earnest.

Far left: An automatic camera on the launch-pad gantry looks up into the Shuttle's main engines and the fumes from the solid rocket boosters during launch.

THE INTERIOR OF THE SHUTTLE ORBITER

The Space Shuttle is designed to carry up to seven people into orbit. Two of these are the mission commander and pilot; the others are technical and scientific personnel. Flying to and from space aboard the Shuttle Orbiter does not impose as great a strain on the flight crew as previous space missions. Acceleration during lift-off is limited to about 3g (three Earth gravities) and re-entry forces are normally less than 1.5g. For this reason personnel in good health can travel with only a minimum of pre-flight training. The Orbiter cabin is designed as a combined working, living and storage area. Seating for four crewmen is provided on the flight deck; the mid-deck has provision for three more crewmen. Beneath this is an equipment and stowage bay reached through an airlock.

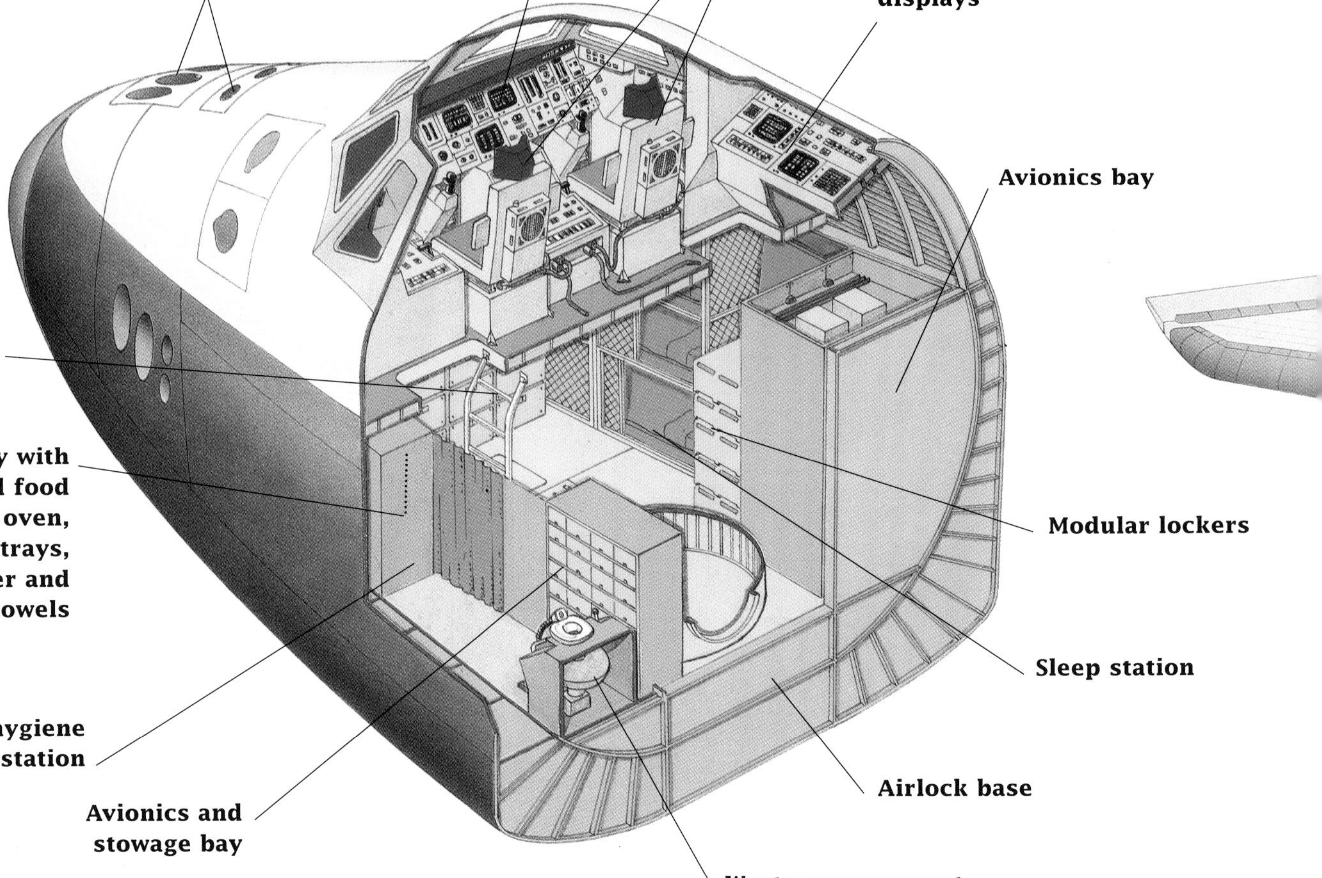

Flight deck control and displays

Seats for commander (left) and pilot (right)

Mission station controls and displays

Forward control thrusters

Avionics bay

Access to flight deck

Galley with refrigerated food store, oven, eating trays, drinks, water and hand towels

Modular lockers

Sleep station

Personal hygiene station

Airlock base

Avionics and stowage bay

Waste management compartment

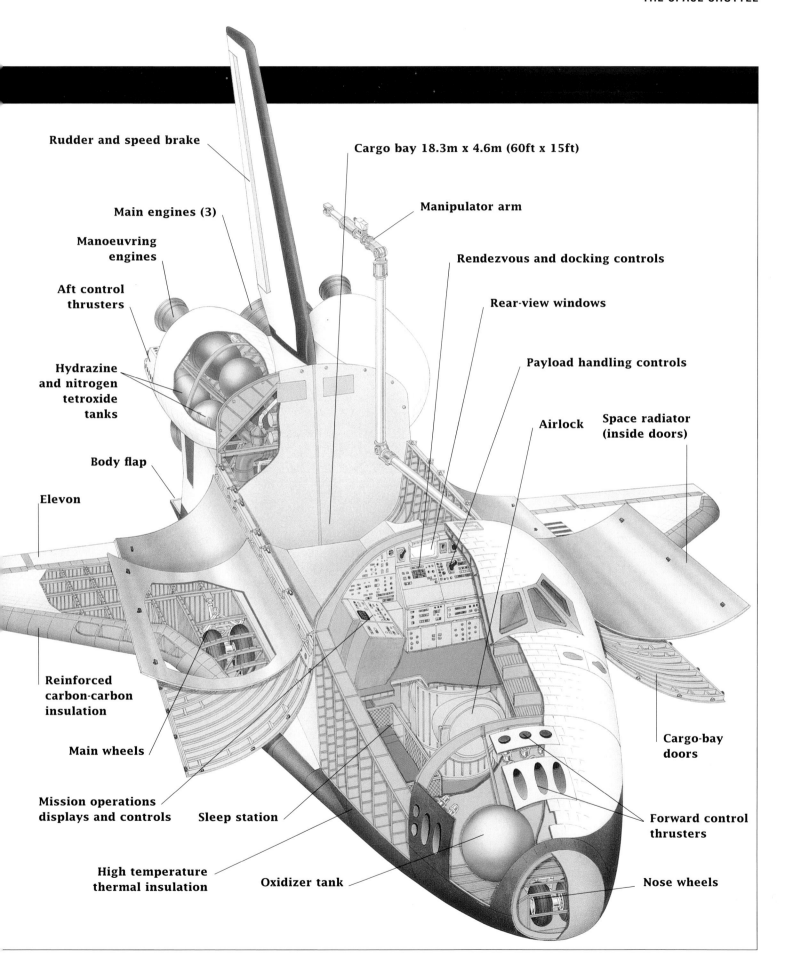

Rudder and speed brake

Cargo bay 18.3m x 4.6m (60ft x 15ft)

Main engines (3)

Manipulator arm

Manoeuvring engines

Rendezvous and docking controls

Aft control thrusters

Rear-view windows

Hydrazine and nitrogen tetroxide tanks

Payload handling controls

Airlock

Space radiator (inside doors)

Body flap

Elevon

Cargo-bay doors

Reinforced carbon-carbon insulation

Main wheels

Forward control thrusters

Mission operations displays and controls

Sleep station

High temperature thermal insulation

Oxidizer tank

Nose wheels

THE CHALLENGER ACCIDENT

In order to continue funding the Space Shuttle during the inevitable delays in its development, which caused the postponement of the first launch from 1978 to 1981, NASA continued to exaggerate the project, claiming that it would make space flights routine and safe. Many people believed the hype and, although the Shuttle never achieved a flight rate as advertised, it made some extraordinary accomplishments, including the capture, repair and return to orbit, or Earth, of malfunctioning satellites. Some flights flew non-astronaut scientists. Soon, there was the demand for other "ordinary" people to fly on the Shuttle. Keen to keep the interest and funds going, NASA flew politicians on the Shuttle as "passenger observers." Later, a teacher, Christa McAuliffe, was selected to fly, after winning a national competition; even journalists were waiting in the space queue. The hype and frenzy seemed to be getting out of hand. The Shuttle system was still in the equivalent of a test-flight phase but was being advertised as a vehicle safe and routine enough to carry passengers. On the 25th flight of the Shuttle, by the orbiter Challenger, McAuliffe and her six astronaut companions, including one industry scientist, were blasted off on a cold sunny day at Kennedy Space Centre, watched live on TV across the US, and even in classrooms. Seventy-three seconds later, the Shuttle broke apart and exploded, killing the crew. One of the SRBs had sprung a leak. The shock was felt worldwide but it should not have been. The NASA PR machine failed to remind the public that the Space Shuttle was a test-plane and that space flight was dangerous. The Challenger accident was a watershed in space development, and safety became the prime concern thereafter.

Above: The Space Shuttle Challenger disintegrated and exploded 73 seconds after launch, while the solid rocket boosters continued to fire, before they were deliberately destroyed by the range safety officer.

Left: Astronaut Curt Brown operates the Shuttle's thrusters using a controller in the aft cockpit.

Below: A wide-angle view of the Space Shuttle payload bay during a spacewalk by two crewmen, also featuring the Remote Manipulator System (RMS) robot arm.

gimballed for pitch, yaw and roll control, and can be throttled within a range of 67–109 per cent rated thrust, with 100 per cent thrust being the equivalent to 170 tonnes at sea level. The thrust at lift-off should be 100 per cent and builds up to a maximum of 109 per cent about 6.5 seconds after lift-off. The level is reduced during Max Q at about T+60 seconds, and later the engines throttle up again. The orange sparks that appear near the engines after about T-10 seconds and just before the point called, "go for main engine start," result from the burning of gaseous hydrogen in the air near the engines. The SSMEs are also re-usable and have been improved significantly during the lifetime of the Shuttle programme to reduce wear and tear on components, especially those in the turbopumps, such as turbine blades.

A HIGHLY MANOEUVRABLE CRAFT

After SSME shutdown, two Orbital Manoeuvring System (OMS) engines, mounted on pods on either side of the tail,

MAJOR SPACE SHUTTLE MISSIONS

Date	Shuttle	Duration
12 April 1981	Columbia STS 1	2 days 6hr 20min. Maiden flight of Space Shuttle by John Young and Bob Crippen
11 November 1982	Columbia STS 5	5 days 2hr 14min. First commercial mission of Shuttle; deploying two communications satellites.
3 February 1984	Challenger STS 41B	7 days 23hr 15min. First independent space walk using manned manoeuvring unit (MMU); first space mission to end at launch site.
6 April 1984	Challenger STS 41C	6 days 23hr 40min. Captures, repairs and redeploys Solar Max satellite.
8 November 1984	Discovery STS 51A	7 days 23hr 45min. Retrieves lost communications satellites and returns them to Earth.
27 August 1985	Discovery STS 51I	7 days 2hr 14min. Captures, repairs and redeploys Leasat 3 satellite.
30 October 1985	Challenger STS 61A	7 days 0hr 44min. First mission with record eight crew.
28 January 1986	Challenger STS 51L	1min 13sec. Explodes at 14.33km (8.9 miles); Dick Scobee, Mike Smith, Judith Resnik, Ronald McNair, Ellison Onizuka, Christa McAuliffe, Gregory Jarvis are killed; first flight to take off but not reach space; first American in-flight fatalities.
29 September 1988	Discovery STS 26	4 days 1hr 0min. US's return to space two years eight months after Challenger.
4 May 1989	Atlantis STS 30	4 days 0hr 57min. Deploys Magellan for its journey to orbit planet Venus. The first deployment of planetary spacecraft from a manned spacecraft.
24 April 1990	Discovery STS 31	5 days 1hr 16min. Deploys Hubble Space Telescope (HST).
7 May 1992	Endeavour STS 49	8 days 21hr 17min. Retrieves Intelsat 6 and re-boosts it into geostationary orbit. Record-breaking 8hr 29min EVA.
2 December 1993	Endeavour STS 61	10 days 19hr 58min. HST servicing and repair mission. US record of five EVAs.
27 June 1995	Atlantis STS 71	9 days 19hr 23min. One hundredth US manned launch docks with Russian Mir space station and switches crews.
19 November 1996	Columbia STS 80	17 days 15hr 54min. Longest Space Shuttle mission.
29 October 1998	STS 95 Discovery	8 days 21hr 43min. Return to space of Mercury astronaut John Glenn, aged 77, 36 years after his Friendship 7 mission. The oldest person in space.
4 December 1998	STS 88 Endeavour	11 days 19hr 18min. First International Space Station (ISS) assembly mission to join Russian Zarya module with US Unity 1 Node.
23 July 1999	STS 93 Columbia	4 days 22hr 50min. Carries 22.58 tonne record payload and flies with first female commander, Eileen Collins.

Right: The Space Shuttle Columbia lands at Edwards Air Force Base, California, at the end of the first Shuttle mission.

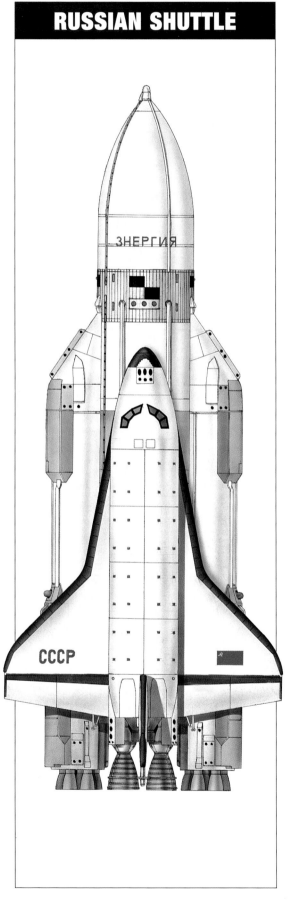

RUSSIAN SHUTTLE

ЗНЕРГИЯ

CCCP

are fired to accelerate the spacecraft to orbital velocity of 7.74km per second (4.81 miles per second). Several firings may be needed to achieve the operational orbit. More recently, the OMS have been used briefly during the ascent to complement the SSMEs. Each launch has its own very specific parameters and requirements. The OMS engines are also the orbiter's retrorockets, which reduce the orbital speed by 91 metres per second (299 feet per second) during a two-minute burn. Each 1.83 tonne thrust of OMS uses hypergolic nitrogen tetroxide and hydrazine propellants which ignite spontaneously on contact.

Yaw, pitch and roll manoeuvres and small velocity changes are performed by the orbiter's Reaction Control System (RCS). The RCS comprises 38 primary thrusters, each with 17,631kg (38,870lb) thrust, of which 14 are on the nose and 12 on each OMS pod, and also six 5.44kg (12lb) thrust verniers, two in the nose and four at the rear, which are powered by similar hypergolic propellants. The RCS engines can be used in an emergency if the OMS engines are not working, for example for a retrofire, as they draw on OMS fuel.

Flight control is regulated by the Shuttle's computer system. Four computers operate in parallel during critical flight operations such as launch, ascent, de-orbit, re-entry and landing, "voting" on every input and response 440 times a second as a safeguard against an individual computer failure. A fifth computer is used as a back-up flight control system. Individual IBM computers of the 1970s class are being replaced by more up-to-date units, and glass-cockpit displays are being introduced to eventually replace the original switches and dials on all orbiters. Pilot astronauts joining the Shuttle programme more recently, having been trained on jets with glass cockpits, find they are faced with the old-fashioned cockpit in the actual Shuttle! The first glass-cockpit mission, by Atlantis, took place in May 2000.

Left: The Russian Space Shuttle Buran flew just one space mission. Launched on the Energia booster, Buran made an automatic orbital flight, landing back at Baikonur Cosmodrome in 1988. The Shuttle programme, along with Energia, was cancelled due to budgetary difficulties.

Above: A Space Shuttle Spacelab laboratory is seen inside the payload bay during a scientific mission crewed by seven astronauts.

A ROBOTIC "LIMB"

Most flights of the Space Shuttle carry the Remote Manipulator System (RMS) a sophisticated robot arm which is controlled from the flight deck by specialist RMS operator astronauts. Positioned in the rear of the cockpit, an operator has control of the RMS components, which include its "hand", a computer and TV cameras, as well as having a view out of the rear window of the flight deck. The RMS arm, 15.24m

(50ft) long, is fitted with movable joints – "shoulder," "elbow," "wrist" – which enable it to be moved into all kinds of positions. It is used to deploy and retrieve payloads, and as a mobile carrier for spacewalking astronauts who stand on a foot restraint at the end of the arm during EVAs. The RMS has proved invaluable to the Shuttle programme, and its technology is also being used in the International Space Station.

COMING BACK DOWN TO EARTH

At the end of a mission, after the OMS retrofire, the Shuttle begins its transition from a spaceship into a glider. It starts re-entry at an altitude of about 122km (75.8 miles) and begins to glide a distance of approximately 8000km (4971 miles) to its landing site at a speed of about Mach 25. As it returns to Earth, the Shuttle orbiter is capable of making manoeuvres during re-entry to enable it to fly 1200km (745.6 miles) to either the left or the right and align itself for an emergency landing. This manoeuvre is called the crossrange. Even

a routine landing will entail some crossrange manoeuvres. The heat builds up due to the friction against the atmosphere, and unlike ablative heat shields on previous manned spacecraft, the Shuttle is equipped with unique thermal tiles and blankets. The thermal protection system (TPS) comprises ceramic and carbon-carbon materials which are applied to the aluminium structure of the orbiter. Like the elements of a gas fire, the TPS tiles glow red when heated, and cool down quickly and undamaged.

Above: The Space Shuttle photographed in orbit by an unmanned free-flying satellite it has deployed. The satellite was later retrieved by the Shuttle and returned to Earth.

Left: Inside a Spacelab laboratory aboard the Space Shuttle, astronauts are hard at work operating a suite of experiments.

NUMBER OF SHUTTLE MISSIONS

Year	Missions
1981	2
1982	3
1983	4
1984	5
1985	9
1986	1 + 1 failure
1987	none
1988	2
1989	5
1990	6
1991	6
1992	8
1993	7
1994	7
1995	7
1996	7
1997	8
1998	5
1999	3
2000	2
Total:	**98, including one failure**

NUMBER OF ORBITER FLIGHTS

Successful flights to 11 February 2000

Columbia	26
Challenger	8 + 1 failure
Discovery	27
Atlantis	21
Endeavour	15

Six types of TPS material are required because the various parts of the orbiter experience different heat levels. The leading edges of the wing and the underside bear the brunt of re-entry heating, whereas the upper part of the fuselage is less exposed to heat. Coated Nomex Felt Reusable Surface Insulation (FRSI) protects the top of the payload bay doors, upper fuselage and OMS pods against temperatures of 700°F. Low-temperature Reusable Surface Insulation (LRSI) comprises 99 per cent pure silica tiles to cope with temperatures of 1200°F over the lower portion of the payload bay doors, fuselage sides, upper surfaces of wings, and parts of the OMS pods. Instead of tiles, some areas are covered with Advanced Flexible Reusable Surface Insulation (AFRSI), a quilted silica-fibre blanket. High-temperature Reusable Surface Insulation (HRSI), comprising silica glass and amorphous fibre tiles, is used on the forward fuselage, cockpit and belly, where temperatures reach 2300°F. These HRSI materials have been replaced partially by Fibrous Refractory Composite Insulation (FRCI) material. Reinforced carbon-carbon tiles, which are coloured black, cope with temperatures exceeding 2300°F on the nose cap and leading edges of the wing.

By the time the orbiter's speed reduces to Mach 10, the vertical stabilizer speed brake can be operated, and at Mach 3.5 the rudder is activated. Automatic or pilot-controlled side-slips and S-turns are carried out to dissipate energy. The approach to the landing site, usually at Kennedy Space Centre, is made at an angle seven times steeper and 20 times faster than that at which an airliner flies. The angle of approach is reduced to 14° at 25.3km (15.7 miles), at a speed of about Mach 2.5, 96km (60 miles) away from the runway. At 14.9km (9.1 miles) altitude, the speed falls below Mach 1 and the runway is 40km (25 miles) flying distance away. The microwave landing system comes into operation 12km (7 miles) from the landing site, and a pre-flare manoeuvre reduces the angle of approach to 1.5° for landing. At an altitude of 91.44m (300ft) the undercarriage is deployed. Twenty-two seconds later, the orbiter has to take its one chance of landing on the 4.57km (2.85 mile) strip at KSC. The speed at

SHUTTLE LANDING PROFILE

This diagram shows the precise manoeuvres and
adjustments to speed and altitude which are necessary
to land the Shuttle safely.

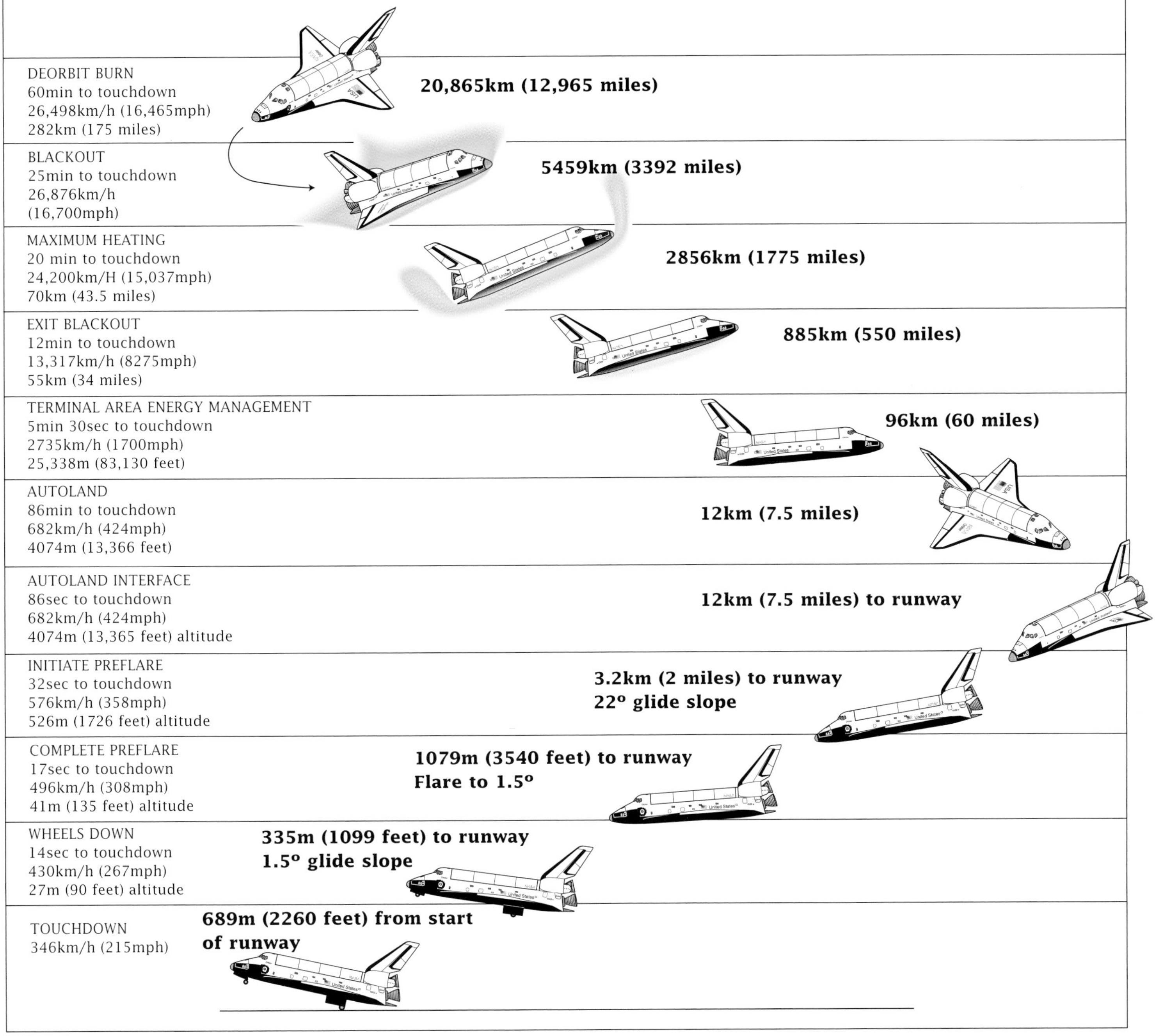

DEORBIT BURN
60min to touchdown
26,498km/h (16,465mph)
282km (175 miles)

20,865km (12,965 miles)

BLACKOUT
25min to touchdown
26,876km/h
(16,700mph)

5459km (3392 miles)

MAXIMUM HEATING
20 min to touchdown
24,200km/H (15,037mph)
70km (43.5 miles)

2856km (1775 miles)

EXIT BLACKOUT
12min to touchdown
13,317km/h (8275mph)
55km (34 miles)

885km (550 miles)

TERMINAL AREA ENERGY MANAGEMENT
5min 30sec to touchdown
2735km/h (1700mph)
25,338m (83,130 feet)

96km (60 miles)

AUTOLAND
86min to touchdown
682km/h (424mph)
4074m (13,366 feet)

12km (7.5 miles)

AUTOLAND INTERFACE
86sec to touchdown
682km/h (424mph)
4074m (13,365 feet) altitude

12km (7.5 miles) to runway

INITIATE PREFLARE
32sec to touchdown
576km/h (358mph)
526m (1726 feet) altitude

3.2km (2 miles) to runway
22° glide slope

COMPLETE PREFLARE
17sec to touchdown
496km/h (308mph)
41m (135 feet) altitude

1079m (3540 feet) to runway
Flare to 1.5°

WHEELS DOWN
14sec to touchdown
430km/h (267mph)
27m (90 feet) altitude

335m (1099 feet) to runway
1.5° glide slope

TOUCHDOWN
346km/h (215mph)

689m (2260 feet) from start
of runway

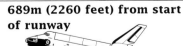

Above: The Space Shuttle's Thermal Protection (TPS) tiles on the underside of the orbiter which receives most of the heat during re-entry.

touchdown is more than 320km/h (199mph). Mission STS 3 set the record with a landing with a landing speed of about 400km/h (249mph) in 1982. The Shuttle owes its landing technique to the many test flights that were flown by small, piloted Shuttle prototypes, called "lifting bodies," such as the X-24. The vehicles were dropped from aircraft to simulate approaches and landings. A drag chute – introduced for flight STS 50 – is deployed to aid breaking. The orbiter is then

142

ABORT!

The first three minutes of launch are the most perilous, especially the two minutes spent on the solid rocket boosters (SRBs). If more than one main engine shuts down during the SRB phase, the stresses on the vehicle, which is then being propelled almost solely by the SRBs, could destroy it. The worst-case scenario is for all the main engines to flame out on the launch pad just as the SRBs ignite. The vehicle would probably be torn apart by the SRB stresses. If an SRB failed in flight, it would spell disaster and certain death. There is simply no escape during the first two minutes. If the Shuttle went out of control it would probably break apart. If the Shuttle was heading uncontrollably towards inhabited areas inland, it would have to be destroyed automatically. Every crew knows the implications. In dire circumstances the orbiter could be jettisoned but would probably break apart under the aerodynamic forces, and would anyway be caught up in the exhaust plume of the SRBs. If one or more engine was lost – or had been lost during the SRB burn – there would be no way of turning back and no way of continuing. (There are options later.)

It is not just the engines that could go wrong. There could be another major systems failure or the cabin may depressurize. The computer software would punch in with a fully automated "contingency abort", which at best could head the Shuttle for a landing in, for example, Bermuda. Other east coast landing sites stretch from South Carolina to Nova Scotia. Failing to reach land means a ditching at sea. Before this happens the crew would have to bail out – provided the orbiter is not below 3.05km (1.9 miles) or above 6km (3.75 miles). The evacuation must also be made in level flight. An extendible pole has to be used to ensure

that the crew slide downward out of the mid-deck main door so that they do not hit the leading edge of the orbiter. Each crew member has to attach his or her parachute harness to a ring, which helps the crew member to slide down the pole and, once free, open the chute. The crew has to bail out because it cannot survive ditching at sea. The astronauts may train for it but in reality the force of impact would kill them.

Above: If anything goes badly wrong with the Space Shuttle during the first two minutes of flight on the SRBs, there is little that can be done to save the crew.

MODIFIED SHUTTLE

One of the major upgrades to the Space Shuttle will be the introduction of liquid-propellant, "fly-back" winged boosters, as seen in this illustration.

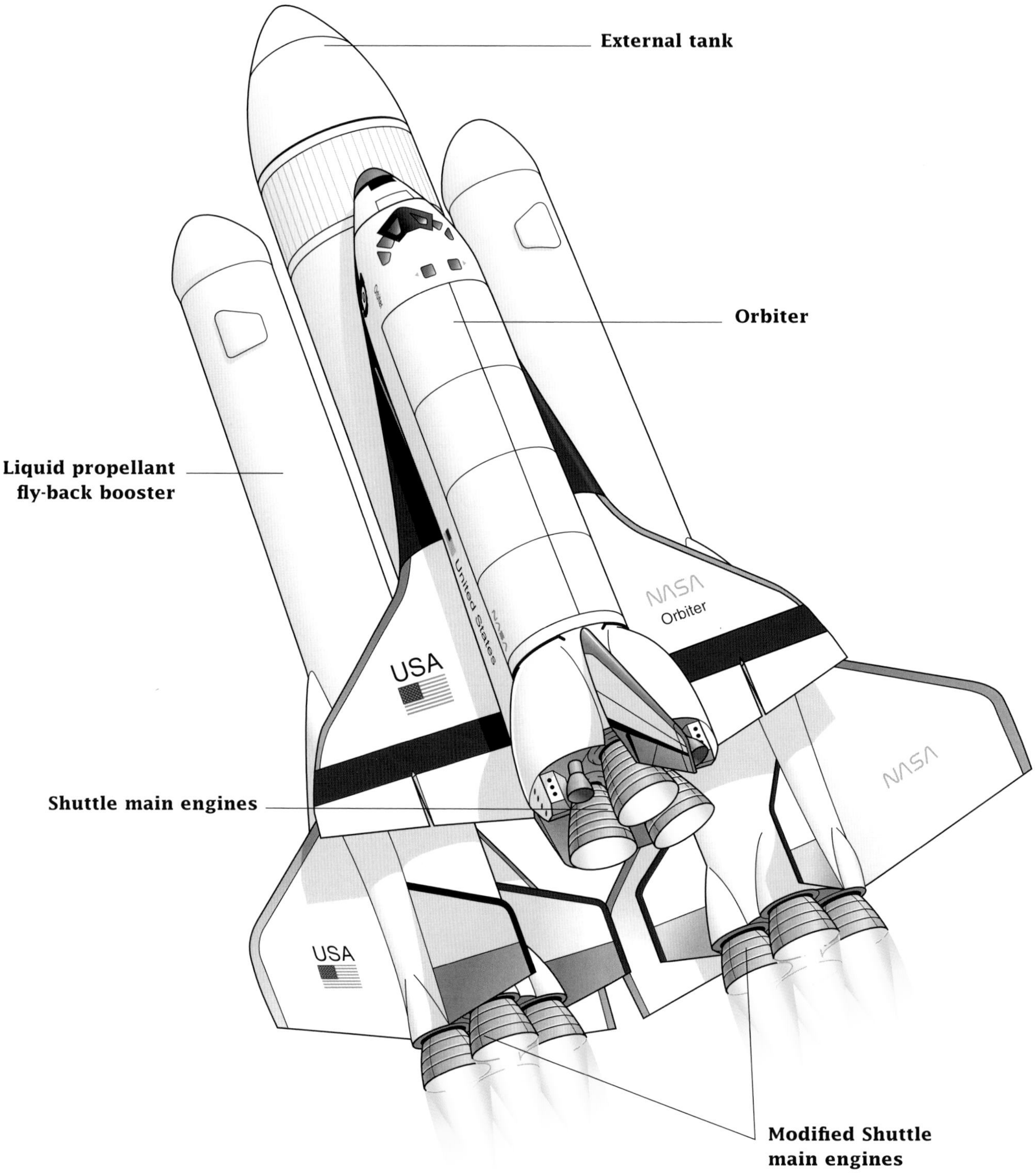

External tank

Orbiter

Liquid propellant fly-back booster

Shuttle main engines

Modified Shuttle main engines

Left: This close-up of Discovery in orbit, taken by an automatic camera on a free-flying satellite, shows the distinctive heat-shield tiles and blankets.

Below: A chase plane guides the Shuttle to its landing at the Kennedy Space Centre, close to the Vehicle Assembly Building in which it was put together for launch.

deactivated and towed back to the Shuttle processing building to be prepared for its next mission. It is transported to the huge Vehicle Assembly Building (VAB) for mating with a new set of SRBs and an ET, before being rolled out to one of the two launch pads, 39A or B, at KSC, ready for its next cataclysmic launch.

SPACE STATIONS

Space stations shaped like bicycle wheels, rotating to create their own gravity. Spaceships taking off like airliners to deliver passengers. These were the images of the Space Age reproduced in books and films in the early days, before the Space Age began. Those dreams have not come true – yet. Space is still an exclusive place for the few to visit.

The real space stations of today are, in reality, a jumble of large tubes, some joined together to form modular platforms. Living in relatively spartan conditions, space station crews are often reminded of the dangerous environment in which they live and the risks involved in travelling to and from orbit. The monumental cost of travelling into space and building space-station components has been a major stumbling block in their development, a point well illustrated by the International Space Station (ISS) which will not be completed until 2004 – ten years later than planned, and at a cost almost ten times that agreed when the project was initiated in 1984. The most famous space station is Russia's Mir, a complex of interlinked modules, the first of which

Left: The Russian Mir Space Station in orbit around Earth. At its centre is the first core module, launched in 1986. Mir will continue to operate into the early years of the 21st century.

was launched in 1986. Mir was still operating in 2000 with plans for it to be de-orbited placed on hold. Mir was preceded into space by a series of Salyut space stations, which consisted mainly of single modules, the first of which was launched in 1971. Compared with Russia, which has amassed huge experience in space-station operations over a period of 30 years, the US has built only one space station, launched in 1973 and manned for less than a year. It was called the Apollo Applications Programme (AAP).

THE US'S SKYLAB

In 1966, as plans were progressing for the Apollo manned missions to the Moon, NASA was looking towards the establishment of a space station in Earth orbit using components developed for Apollo. AAP was not to be NASA's final space station design,

but an interim measure to demonstrate its technology. AAP would be flown during the final Apollo flights to the Moon, creating for NASA a twin "jamboree" of prestigious space exploration. The space station, which was to become known later as Skylab, would be launched using the Saturn V, and astronaut teams would be flown to it in Apollo command and service modules (CSMs) using Saturn 1B boosters.

The space station would comprise a fully equipped Saturn V third stage. The stage, emptied of its propellants, would be converted into a fully-equipped space station by visiting astronauts. This "wet workshop" mode was later changed to a "dry workshop", meaning that the third stage would be completely equipped before launch. Skylab would also carry a telescope which was based on the lunar module. To accommodate Skylab financially, other

Below: The final Saturn V booster lifts Skylab into orbit from the Kennedy Space Centre in May 1973.

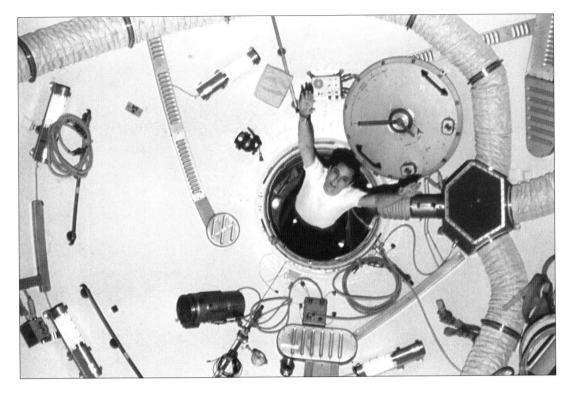

budgets were reduced and later Apollo missions cancelled.

The main part of Skylab was the Orbital Workshop which was based around a Saturn V third stage. The Orbital Workshop was 14.6m (48ft) long and 6.7m (22ft) in diameter, and provided 283 square metres (3046 square feet) of living and work space on two levels, divided by a grid floor. The lower level had the wardroom, kitchen, washroom and living quarters, and a large picture window for viewing Earth. Medical equipment was located on the lower level, and there was also an airlock in the floor containing a tank in which to deposit rubbish. The upper level contained most of the crews' working areas and consumables, such as equipment, personal kit and clothes.

At the top of the Orbital Workshop was access to an airlock module from where space walks could be made. This airlock led to the Multiple Docking Adaptor which enabled more than one CSM to join with the station, for example, in the event of a space rescue being mounted for a failed CSM. Attached to the side of the Multiple Docking Adaptor was the most important instrument on Skylab, the Apollo Telescope Mount, equipped with five telescopes observing in different wavebands, mainly to study the Sun. Indeed, one of Skylab's major accomplishments was the mass of data collected about Earth's nearest star.

Skylab was launched on 14 May 1973, with the first crew scheduled to follow it into orbit a day later. However, the Skylab launch was not altogether successful. A meteoroid shield on the side of the Skylab was torn off during launch, taking with it one of the space station's two solar panels, and jamming the other. It appeared that the space station would be doomed once it had reached orbit.

The first manned mission, dubbed Skylab 2, was going to be more than a routine shuttle – it would be a full-scale rescue operation. Veteran moon walker Pete Conrad took his rookie crew of two to Skylab and over the course of 28 days, which included a dangerous space walk, converted the station into a workable space base. The 59-day Skylab 3 mission was followed by Skylab 4's record of 84 days. In

The Skylab launch was not altogether successful. A meteoroid shield was torn off during launch, taking with it one of the space station's two solar panels, and jamming the other. It appeared that the space station would be doomed once it reached orbit.

SKYLAB

Skylab was America's first manned space station. Astronauts in Mercury, Gemini and Apollo had lived in cramped quarters, and largely ate pastes and liquids out of bags. For the first time, Skylab offered some of the more common creature comforts. There was water for occasional showers, and food stowed in containers and freezers which offered a more varied diet. Clothing was stored in "28-day clothing modules" in lockers; nothing was washed in the space station, and used clothes were disposed of in an empty tank beneath the floor of the living quarters. The biggest luxury was "waste management compartment" which, while serving the needs of the crew, was really a medical laboratory allowing study of the astronaut's mineral and bodily fluids balance.

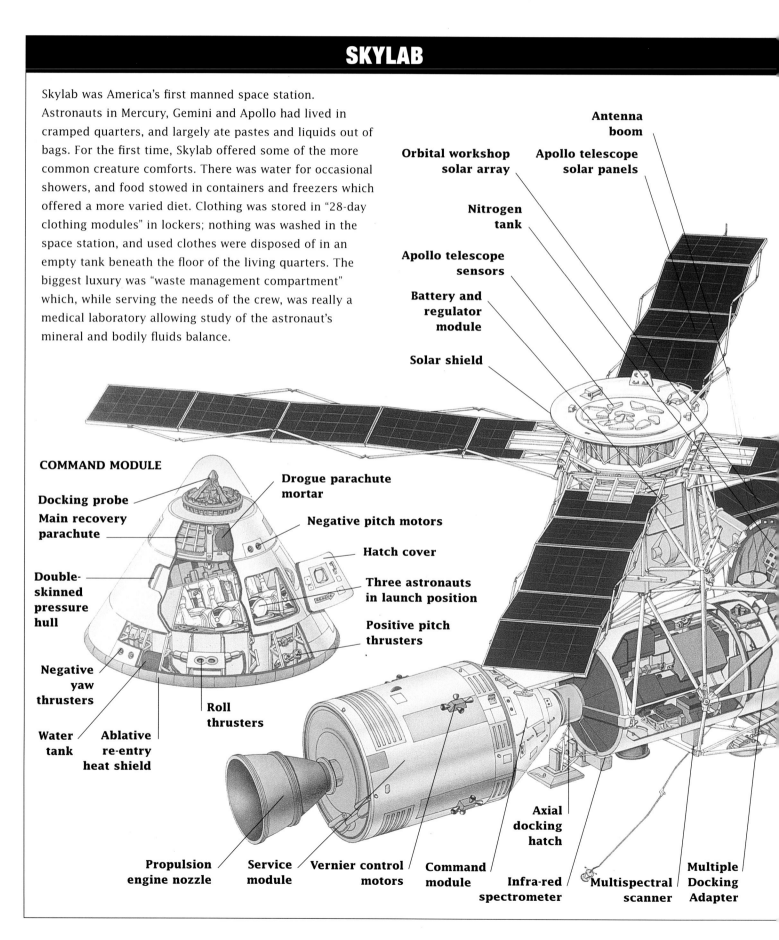

Antenna boom

Orbital workshop solar array

Apollo telescope solar panels

Nitrogen tank

Apollo telescope sensors

Battery and regulator module

Solar shield

COMMAND MODULE

Docking probe

Main recovery parachute

Drogue parachute mortar

Negative pitch motors

Hatch cover

Three astronauts in launch position

Positive pitch thrusters

Double-skinned pressure hull

Negative yaw thrusters

Roll thrusters

Water tank

Ablative re-entry heat shield

Propulsion engine nozzle

Service module

Vernier control motors

Command module

Infra-red spectrometer

Axial docking hatch

Multispectral scanner

Multiple Docking Adapter

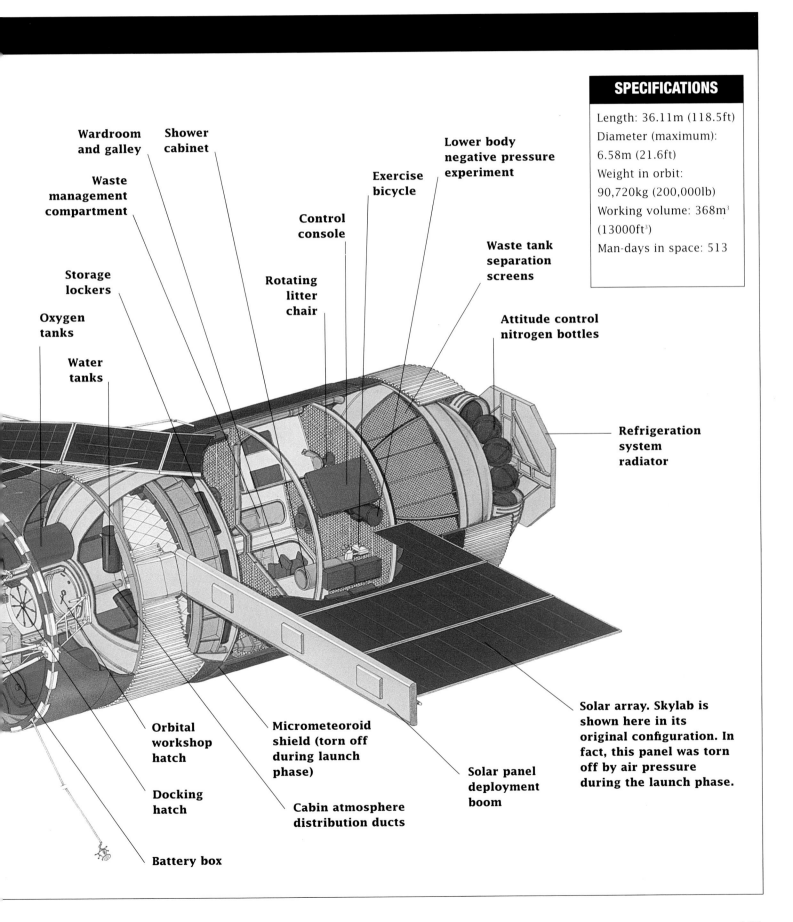

Wardroom and galley

Shower cabinet

Waste management compartment

Lower body negative pressure experiment

Exercise bicycle

Control console

Storage lockers

Rotating litter chair

Oxygen tanks

Water tanks

Waste tank separation screens

Attitude control nitrogen bottles

Refrigeration system radiator

Orbital workshop hatch

Micrometeoroid shield (torn off during launch phase)

Solar panel deployment boom

Solar array. Skylab is shown here in its original configuration. In fact, this panel was torn off by air pressure during the launch phase.

Docking hatch

Cabin atmosphere distribution ducts

Battery box

SPECIFICATIONS

Length: 36.11m (118.5ft)
Diameter (maximum):
6.58m (21.6ft)
Weight in orbit:
90,720kg (200,000lb)
Working volume: 368m^3
(13000ft^3)
Man-days in space: 513

SOYUZ-T

The Soyuz-T space station ferry was revealed by the Soviets at the end of 1979 and first flew with men the following year. Although it retained the basic Soyuz design it had been greatly modified. The Soviets proudly announced that the Soyuz-T spacecraft carried a new, improved computer system, but this seemed only to match the capabilities which American manned spacecraft possessed more than a decade previously. Although Soyuz-T could carry three people, many of the crews comprised only two. Soyuz-T became operational with the Salyut 7 missions.

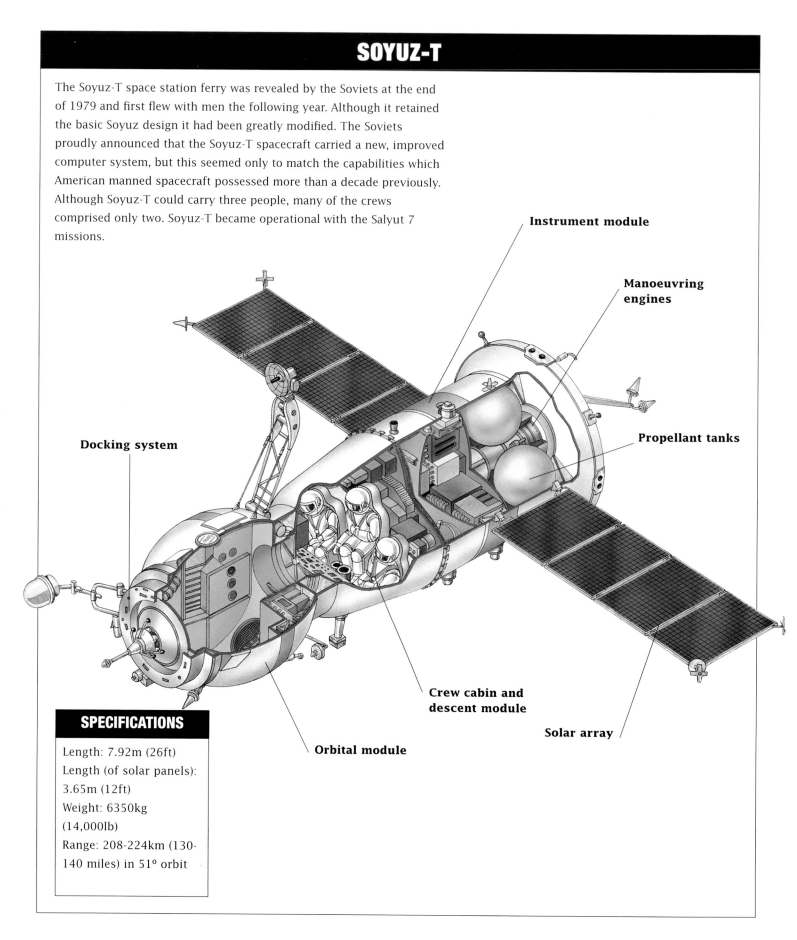

Instrument module

Manoeuvring engines

Propellant tanks

Docking system

Crew cabin and descent module

Solar array

Orbital module

SPECIFICATIONS

Length: 7.92m (26ft)
Length (of solar panels):
3.65m (12ft)
Weight: 6350kg
(14,000lb)
Range: 208-224km (130-
140 miles) in 51° orbit

SALYUT 1

Salyut 1 is seen attached to the Soyuz space ferry. The space station has two pairs of solar panels and a main laboratory area.

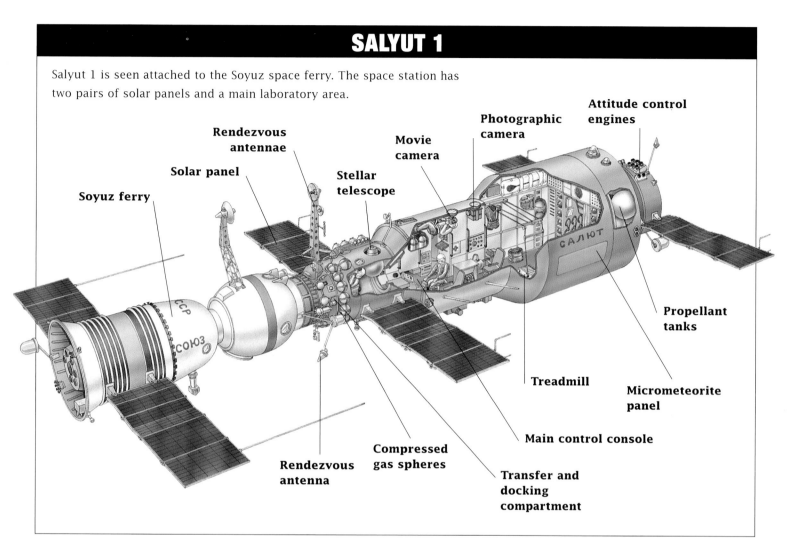

Attitude control engines

Photographic camera

Rendezvous antennae

Movie camera

Solar panel

Stellar telescope

Soyuz ferry

Propellant tanks

Treadmill

Micrometeorite panel

Main control console

Rendezvous antenna

Compressed gas spheres

Transfer and docking compartment

hindsight, Skylab was an immense scientific achievement. It is a great pity that it was abandoned in orbit for the next five years before it made a spectacular re-entry, showering parts of the Australian outback with small pieces of debris.

THE SOVIET UNION'S SPACE FERRY

The Soviet Union launched the first space station, called Salyut, two years before Skylab, in 1971. Salyut space-station plans first depended on the development of a ferry vehicle to carry crews to and from space. This was called the Soyuz, which was capable of flying three cosmonauts to the Salyut stations, and also of flying independent research flights in orbit.

Soyuz weighed about 6.6 tonnes and was about 8.35m (27.4ft) long. It comprised an instrument module, which contained a propulsion system used for retrofire and in-orbit manoeuvres; the descent module, which was the flight capsule; and an orbital module, which was a pressurized cabin that could be used for experiments on independent flights and, as a docking unit, if required. The span of the solar array either side of the instrument section was 9.5m (31.2ft).

The first manned Soyuz was flown on 23 April 1967 in an unnecessarily risky flight designed to win something out of the Space Race. The plan was to dock two vehicles in orbit and transfer crews from one to the other, something that the US Gemini had not achieved. The flight was a disaster and the second launch cancelled. The lone Soyuz 1 cosmonaut, Vladimir Komarov, was

APOLLO-SOYUZ TEST PROJECT

The world's most ambitious experiment in space co-operation took place in July 1975 when an American Apollo spacecraft docked with a Russian Soyuz capsule 225km (140 miles) above the Earth. The mission, known as the Apollo-Soyuz Test Project (ASTP), in which astronauts and cosmonauts entered one another's spacecraft and conducted experiments, was wholly successful. To join Apollo and Soyuz, which had different pressures, it was necessary for the Americans to take into orbit a docking module jointly developed by American and Russian engineers. The cosmonauts and astronauts entered here to acclimatize to the different pressures before entering the other ship.

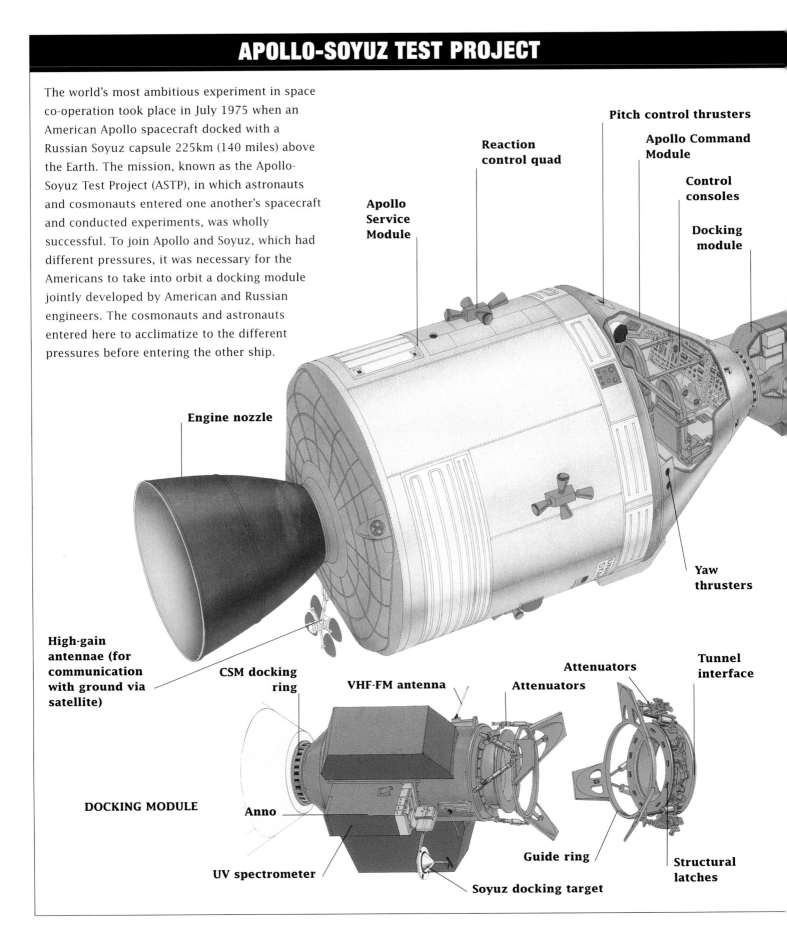

Pitch control thrusters

Reaction control quad

Apollo Command Module

Control consoles

Apollo Service Module

Docking module

Engine nozzle

Yaw thrusters

High-gain antennae (for communication with ground via satellite)

CSM docking ring

VHF-FM antenna

Attenuators

Attenuators

Tunnel interface

DOCKING MODULE

Anno

UV spectrometer

Soyuz docking target

Guide ring

Structural latches

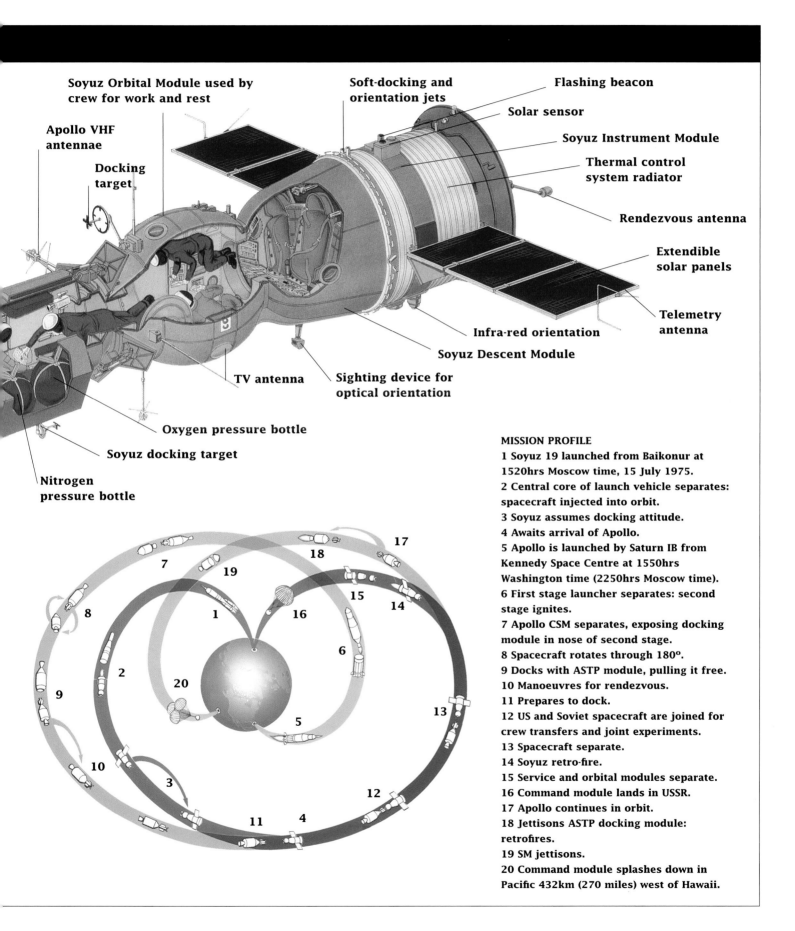

Soyuz Orbital Module used by crew for work and rest

Apollo VHF antennae

Docking target

Soft-docking and orientation jets

Flashing beacon

Solar sensor

Soyuz Instrument Module

Thermal control system radiator

Rendezvous antenna

Extendible solar panels

Telemetry antenna

Infra-red orientation

Soyuz Descent Module

TV antenna

Sighting device for optical orientation

Oxygen pressure bottle

Soyuz docking target

Nitrogen pressure bottle

MISSION PROFILE
1 Soyuz 19 launched from Baikonur at 1520hrs Moscow time, 15 July 1975.
2 Central core of launch vehicle separates: spacecraft injected into orbit.
3 Soyuz assumes docking attitude.
4 Awaits arrival of Apollo.
5 Apollo is launched by Saturn IB from Kennedy Space Centre at 1550hrs Washington time (2250hrs Moscow time).
6 First stage launcher separates: second stage ignites.
7 Apollo CSM separates, exposing docking module in nose of second stage.
8 Spacecraft rotates through 180°.
9 Docks with ASTP module, pulling it free.
10 Manoeuvres for rendezvous.
11 Prepares to dock.
12 US and Soviet spacecraft are joined for crew transfers and joint experiments.
13 Spacecraft separate.
14 Soyuz retro-fire.
15 Service and orbital modules separate.
16 Command module lands in USSR.
17 Apollo continues in orbit.
18 Jettisons ASTP docking module: retrofires.
19 SM jettisons.
20 Command module splashes down in Pacific 432km (270 miles) west of Hawaii.

killed when the parachute on his spacecraft failed to open. Other Soyuz flights were made, including an unusual three-craft mission in 1969. One unique Soyuz was used in 1975 for the docking of Soviet and Apollo spacecraft in Earth orbit. The Apollo-Soyuz project was a fine demonstration of space co-operation, but it was not followed up. Soyuz 19 was launched with two cosmonauts, including veteran space walker Alexei Leonov, and was followed by Apollo 18 with a three-person crew.

The first Soyuz flown as a space-station ferry was not launched until 1971. Named Soyuz 10, it failed to make a hard dock at Salyut 1, and it was left to the three-man crew of Soyuz 11 to make a successful docking and transfer into the orbital base. All three cosmonauts were killed when their spacecraft depressurized before re-entry, after a stay lasting 23 days, and the folly of their wearing tracksuits rather than spacesuits was exposed.

A new form of Soyuz space station ferry was later developed. Instead of solar panels, the ferry was simply equipped with batteries to provide the necessary power for the two-day flight to the station and return to Earth. It first flew on an independent manned test flight in 1973. After several docking failures with Salyut space stations, necessitating the immediate return of the crew to Earth, all Soyuz craft were equipped with solar panels once again.

SALYUT 1

The design of Salyut 1 was retained in a series of subsequent stations, ending with Salyut 7. Even the new space station, Mir, owed its design to Salyut 1. The first Salyut space station was launched on 19 April 1971. The station weighed nearly 19 tonnes and was 14.5m (47.6ft long). It was equipped with two pairs of solar panels, one fixed to the rear instrument compartment, and another pair fixed to the forward of three workstations. The docking and transfer section, built to receive the Soyuz ferry and its crew, led to the first workstation which was 2.9m (9.5ft) in diameter and 3.8m (12.47ft) long, and which in turn led to the largest work compartment, 4.1m (13.5ft) long and 4.15m (13.60ft) in diameter. At the rear sat the propulsion system, 2.17m (7.1ft) long

SALYUT 4

The design of Salyut 4 was based on that of Salyut 1, the major difference being the replacement of the earlier four non-steerable solar panels with three large steerable panels. A hatch cover was incorporated into the forward work compartment, and this would have allowed the transfer compartment to act as an airlock during spacewalk activities. Although spacewalks were apparently planned, they were cancelled – possibly as a result of the April launch abort and the delay in re-manning the station.

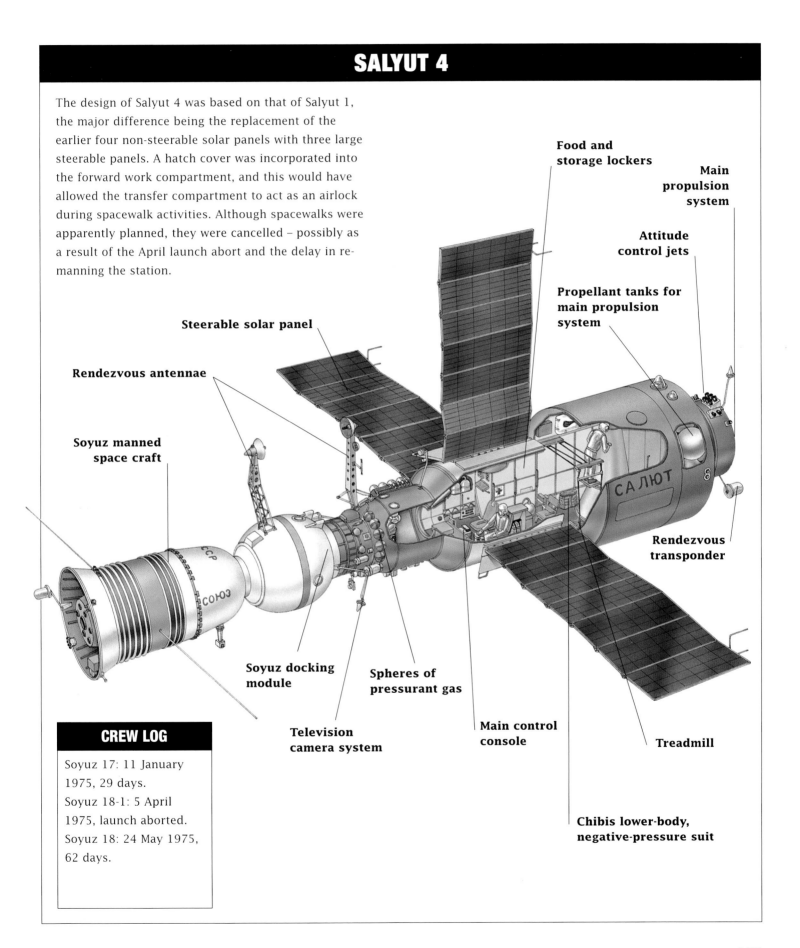

Food and storage lockers

Main propulsion system

Attitude control jets

Propellant tanks for main propulsion system

Steerable solar panel

Rendezvous antennae

Soyuz manned space craft

Rendezvous transponder

Soyuz docking module

Spheres of pressurant gas

Treadmill

Television camera system

Main control console

Chibis lower-body, negative-pressure suit

CREW LOG

Soyuz 17: 11 January 1975, 29 days.
Soyuz 18-1: 5 April 1975, launch aborted.
Soyuz 18: 24 May 1975, 62 days.

and 2.2m (7.2ft) in diameter, which housed the manoeuvring and retrofire engine which had a total burn time of 16 minutes 40 seconds. Salyut carried over 1300 instruments, including telescopes, cameras and sensors for astronomy, science and remote sensing, and exercise equipment to keep the cosmonauts physically fit during their long stay in a weightless environment. Salyut 1 re-entered and burned up in Earth's atmosphere in October 1971, after housing just one crew.

SPIES IN SPACE

In 1973, the Soviets launched Salyut 2. Although details of the station have never been released, evidence suggests that this resembled Salyut 1. The main difference was that the Soyuz ferry vehicle now docked at the rear-propulsion section, which was redesigned to accommodate the docking port. The workstations were basically of the same design, but at the front, there was a conical craft designed to undock and return to Earth. From the sparse Soviet announcements made about Salyut 2, it was clear that it was probably a military space station, equipped with reconnaissance cameras which would return to Earth with the capsule.

Salyut 2 actually failed in orbit and was never manned, but the Salyut 3 and 5 stations followed on 24 June 1974 and 22 June 1976 respectively. These were manned by three two-person crews while two other crews failed to dock. Although some scientific work was conducted, it is clear that much of the work was secret, possibly

PROGRESS

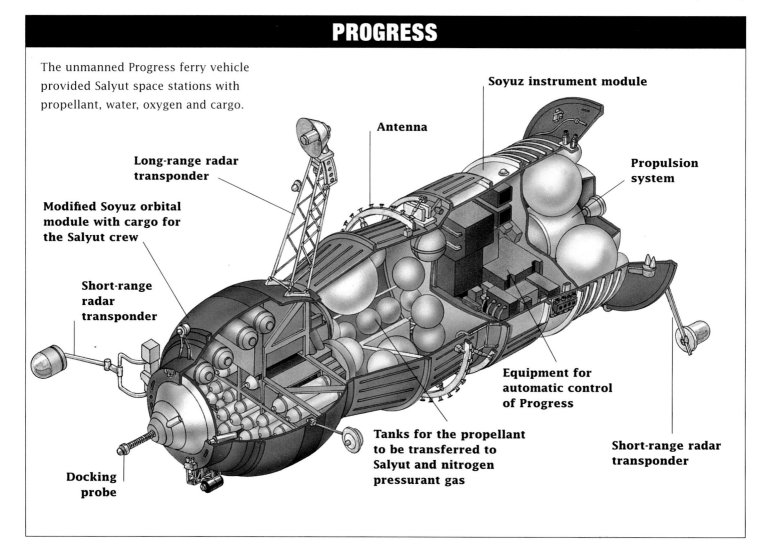

The unmanned Progress ferry vehicle provided Salyut space stations with propellant, water, oxygen and cargo.

Soyuz instrument module

Antenna

Long-range radar transponder

Propulsion system

Modified Soyuz orbital module with cargo for the Salyut crew

Short-range radar transponder

Equipment for automatic control of Progress

Tanks for the propellant to be transferred to Salyut and nitrogen pressurant gas

Short-range radar transponder

Docking probe

SPACE STATION MISSION HIGHLIGHTS

Date	Space Station	Mission
6 June 1971	SALYUT 1	Georgi Dobrovolsky, Vladislav Volkov and Viktor Patsayev make record 23-day flight, but die as Soyuz 11 craft depressurizes before re-entry.
25 May 1973	SKYLAB 2	Three astronauts make 28-day flight with spacewalks to repair severely disabled space station.
28 July 1973	SKYLAB 3	Three astronauts make 59-day mission.
16 November 1973	SKYLAB 4	Three astronauts make record-breaking 84-day mission, the last to Skylab.
3 July 1974	SALYUT 3	Soyuz 14's Pavel Popovich and Yuri Artyukhin are first space spies in 15-day mission.
11 January 1975	SALYUT 4	Two Soyuz 17 cosmonauts make 29-day mission.
5 April 1975	SALYUT 4	Two Soyuz 18-1 cosmonauts, Vasili Lazarev and Oleg Makarov, make first launch abort as second stage fails.
24 May 1975	SALYUT 4	Two Soyuz 18 cosmonauts work for 62 days.
6 July 1976	SALYUT 5	Two Soyuz 21 cosmonauts remain for 49 days but evacuate Salyut 5 after emergency.
10 December 1977	SALYUT 6	Soyuz 26's Yuri Romanenko and Georgi Grechko stay 96 days, breaking Skylab 4 record.
2 March 1978	SALYUT 6	Soyuz 28 makes seven-day visit with two-man crew including Czechoslovakian Vladimir Remek who becomes the first non-American, non-Soviet in space, and the first of several visitors from Soviet bloc countries.
15 June 1978	SALYUT 6	Soyuz 29's Vladimir Kovalyonok and Alexander Ivanchenkov make record 139-day flight.
25 February 1979	SALYUT 6	Soyuz 32's Vladimir Lyakhov and Valeri Ryumin make record 175-day flight.
9 April 1980	SALYUT 6	Soyuz 35's Leonid Popov and Valeri Ryumin make 184-day flight.
5 June 1980	SALYUT 6	Soyuz T2's two-man crew test new Soyuz version during 3-day flight.
27 November 1980	SALYUT 6	Soyuz T3 carries three-man maintenance crew for 12-day flight.
12 March 1981	SALYUT 6	Soyuz T4's two-man crew makes 74-day mission as final Salyut 6 long-stay crew.
15 May 1981	SALYUT 6	Two-man Soyuz 40 crew, making final mission to Salyut, includes a Romanian cosmonaut. Mission lasts seven days.
13 May 1982	SALYUT 7	Soyuz T5's Anatoli Berezevoi, and Valentin Lebedev make record 211-day mission.
27 June 1983	SALYUT 7	Soyuz T9's two-man crew make 149-day mission.
27 September 1983	SALYUT 7	Aboard Soyuz T10-1, Vladimir Titov and Gennadi Strekalov escape by using launch escape system as launcher explodes on pad.
8 February 1984	SALYUT 7	Soyuz T10's Leonid Kizim, Vladimir Solovyov and doctor Oleg Atkov make record-236-day flight, including six spacewalks.
17 July 1984	SALYUT 7	Soyuz T12 has three-person crew including Svetlana Savitskaya who is the first woman to walk in space during 11-day flight.
6 June 1985	SALYUT 7	Soyuz T13 with Vladimir Dzhanibekov and Viktor Savinykh overhaul and repair Salyut 7 after systems failures. Main mission lasts 112 days.

involving the use of a 10m (32.8ft) focal length camera with a 1m (3.3ft) resolution of objects on the ground. The Soviets announced that a recoverable spacecraft had been released by both stations, and had returned to Earth. Salyut 3 re-entered the Earth's atmosphere on 24 January 1975, and Salyut 5 on 8 August two years later.

CIVILIAN SALYUTS

Between the military Salyut 3 and 5 missions, the civilian Salyut 4 was launched on 26 December 1974. The body of the space station was almost identical to Salyut 1, but instead of two pairs of solar panels, Salyut 3 had one set of three, each fixed to a different side of the first workstation. One

SALYUT 6

With the appearance of Salyut 6 in September 1977 the Soviet space station programme became more ambitious. The new station had docking ports fore and aft and the propulsion system was re-arranged to permit refuelling by Progress cargo ships which docked on the aft airlock. Unlike America's Skylab, Salyut 6 had a re-startable rocket engine which could be used to manoeuvre it into a higher orbit. Salyut 6 served the Soviet Union superbly for over three years, and its crews established new duration records. The station also hosted visits from several international crews. This illustration shows Salyut with both Soyuz and Progress docked.

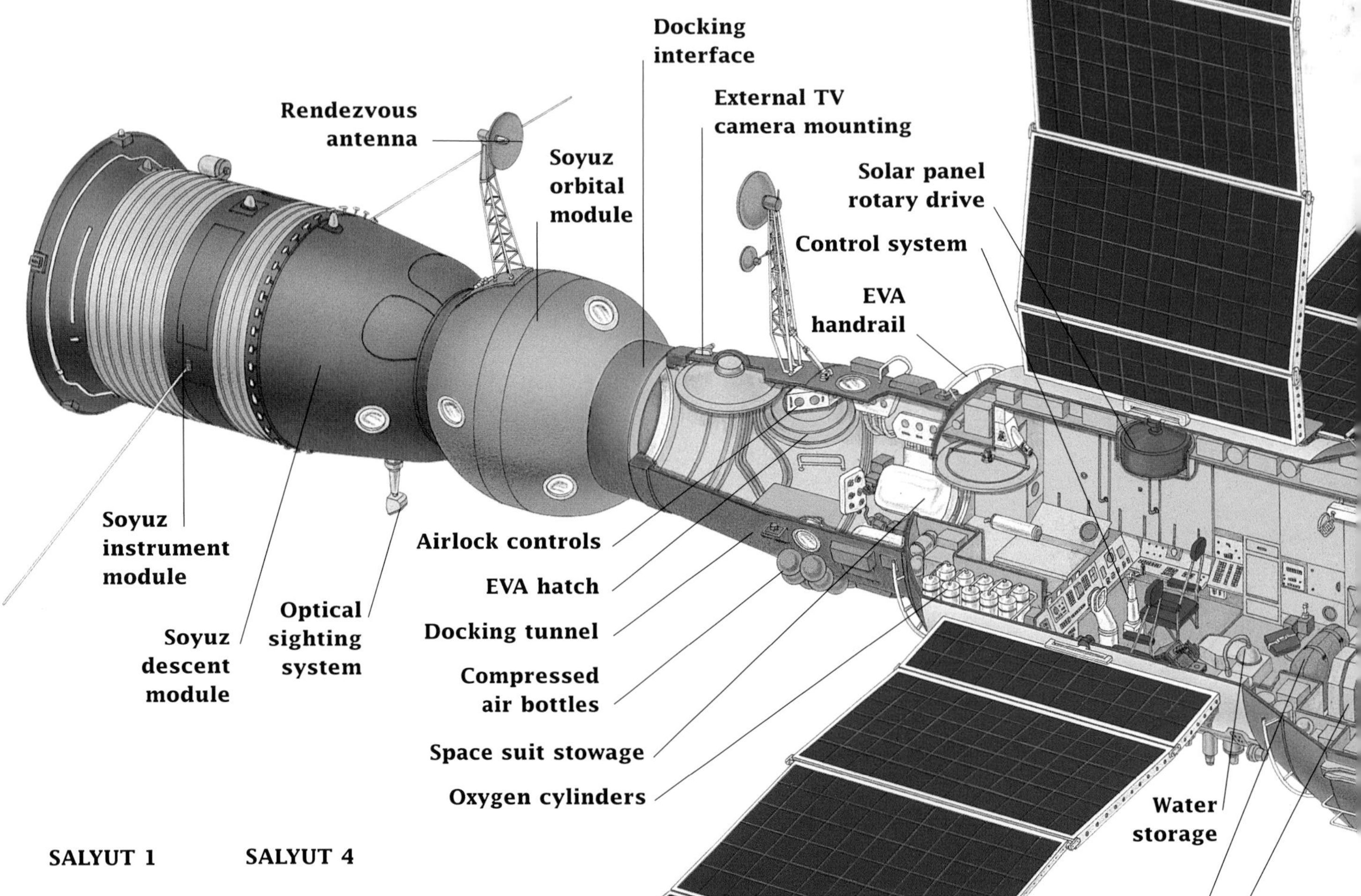

Docking interface

Rendezvous antenna

External TV camera mounting

Soyuz orbital module

Solar panel rotary drive

Control system

EVA handrail

Soyuz instrument module

Airlock controls

EVA hatch

Docking tunnel

Compressed air bottles

Space suit stowage

Oxygen cylinders

Soyuz descent module

Optical sighting system

Water storage

Vacuum cylinder

Multispectral camera

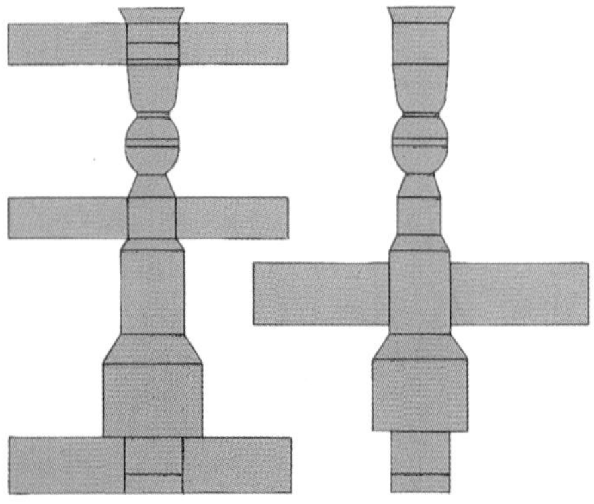

SALYUT 1

SALYUT 4

Left: The original Soyuz ferry with solar panels on Salyut 1, compared with the Salyut 4 ferry which had no solar panels

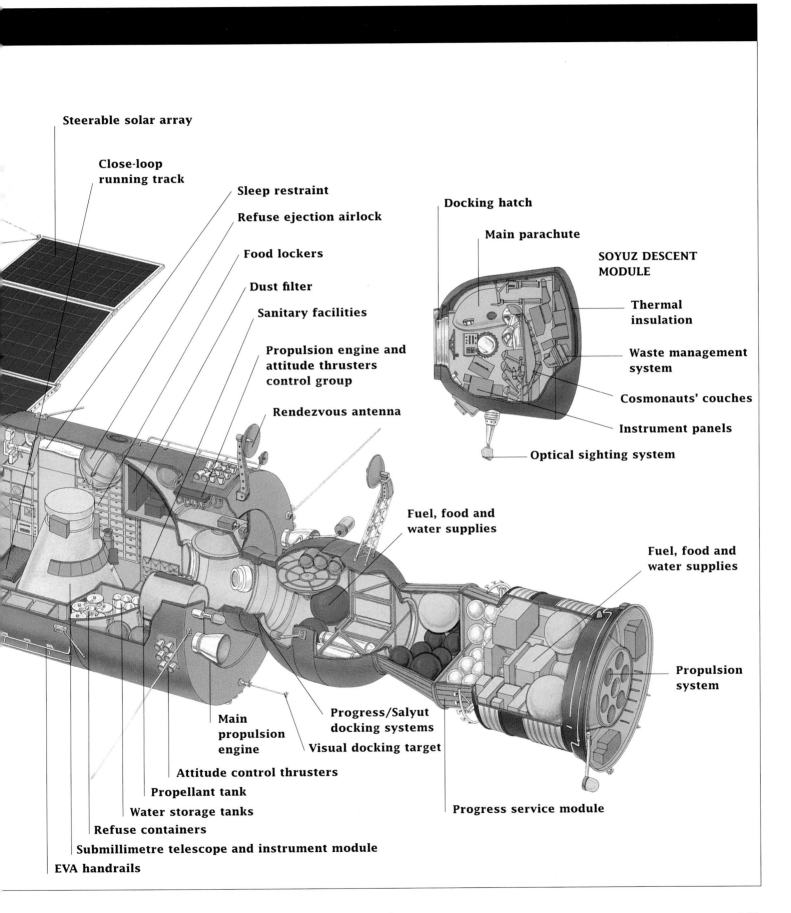

Steerable solar array

Close-loop
running track

Sleep restraint

Refuse ejection airlock

Food lockers

Dust filter

Sanitary facilities

Propulsion engine and
attitude thrusters
control group

Rendezvous antenna

Docking hatch

Main parachute

**SOYUZ DESCENT
MODULE**

Thermal
insulation

Waste management
system

Cosmonauts' couches

Instrument panels

Optical sighting system

Fuel, food and
water supplies

Fuel, food and
water supplies

Propulsion
system

Main
propulsion
engine

Progress/Salyut
docking systems

Visual docking target

Attitude control thrusters

Propellant tank

Water storage tanks

Refuse containers

Submillimetre telescope and instrument module

EVA handrails

Progress service module

SALYUT 7

Salyut 7 is pictured with a Soyuz manned ferry craft (right) and the Cosmos 1443 module (left) which included a re-entry capsule.

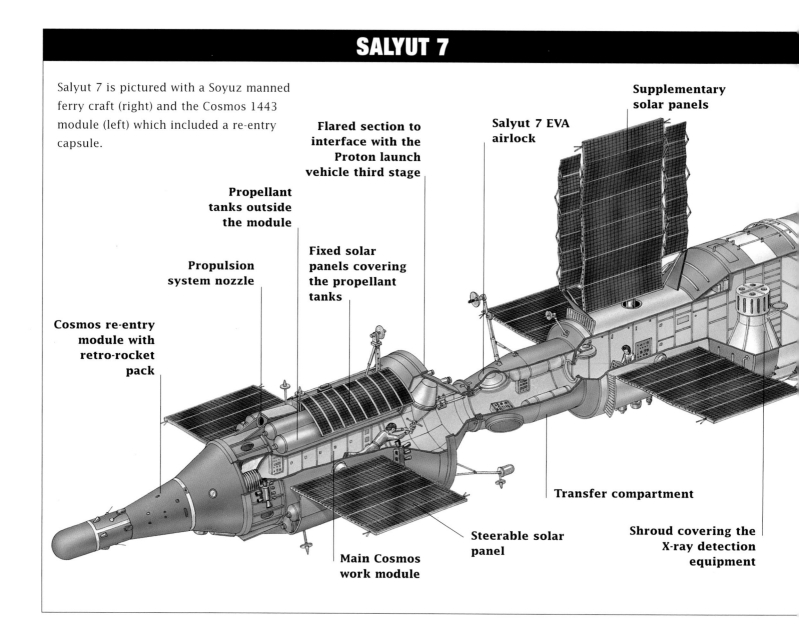

Supplementary solar panels

Flared section to interface with the Proton launch vehicle third stage

Salyut 7 EVA airlock

Propellant tanks outside the module

Propulsion system nozzle

Fixed solar panels covering the propellant tanks

Cosmos re-entry module with retro-rocket pack

Transfer compartment

Steerable solar panel

Shroud covering the X-ray detection equipment

Main Cosmos work module

Far right: Salyut 7 is pictured in orbit. Launched in 1982, the station was last manned in 1985.

of the major instruments on Salyut 4 was the Orbital Solar Telescope (OST) which took up most of the rear section of the larger workstation. Six other astronomical experiments were conducted on the station, along with seven medical experiments, and a series of biological experiments. Two long-duration crews were kept busy, making space-station work seem routine, while one crew was thwarted by a launch failure, resulting in an emergency landing in April 1975. Salyut 4 re-entered Earth's atmosphere on 2 February 1977.

Salyut 6, launched on 29 September 1977, was the first of a second generation

series of Salyuts, although it resembled Salyut 4. One of its major innovations was the introduction of an unmanned Soyuz-class ferry, called Progress, which docked at the rear of the station to provide fuel, water, oxygen, food, post and personal possessions for the numerous crews which operated aboard it. These included an array of visiting cosmonauts from Soviet-bloc countries, flown for propaganda purposes.

Like Salyut 4, the interior of the new station was dominated by a large telescope, called the BST-1M submillimetre telescope, with a liquid nitrogen cryogenic cooling

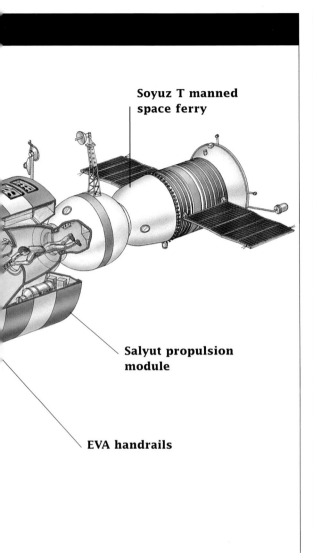

Soyuz T manned space ferry

Salyut propulsion module

EVA handrails

unit. Other major instruments included the KATE-140 stereographic, topographical camera, and the MKF-6M earth resources camera. Regimented work aboard Salyut 6 often involved routine maintenance space walks, which on one occasion involved the disposal of a large radio telescope temporarily deployed at the rear of Salyut. During the lifetime of Salyut 6 , a new improved Soyuz ferry, called the Soyuz T, was introduced as part of an unmanned test flight of Salyut 6. One of the Soyuz T's major innovations was the ability to carry three crewmen instead of two. Another introduction was the use of a new module launched on 25 April 1981 to dock with Salyut 6. Called the Star module, it was designated officially as Cosmos 1267. Before it docked with Salyut, it deployed a re-entry capsule, raising speculation that it was a redundant military-class vehicle recalled into service. Cosmos 1267 was also used to perform the de-orbit burn for Salyut 6 on 29 July 1982, because the station's own propulsion unit was not functioning correctly.

Salyut 7 was launched on 19 April 1982. During its operation, it was joined by two Heavy Cosmos modules, 1443 and 1686. These modules added extra work space and solar power, and 1443 module also released a re-entry capsule while docked at the station. Both modules were almost the same size as Salyut 7. Towards the end of its life, Salyut 7 became severely degraded, and contact with the station was lost in February 1985. In a mission similar to the

Right: A Soyuz booster blasts off from the Baikonur Cosmodrome, carrying a three-man crew on a flight to Mir. A launch escape system tower is located on top of the booster.

one made to restore the Skylab space station to service, two cosmonauts were launched in Soyuz T13 to perform a similar task, and in extremely inhospitable conditions they managed to bring the station back to life. Having done so, Cosmos 1686 was launched in the hope that it would support another manned mission. However, the mission had to be terminated when a cosmonaut became ill, and again Salyut 7 was abandoned. After the first module of a new generation space station, Mir, had been launched, its first crew paid a last visit to Salyut 7 before it made a natural re-entry much later on 7 February 1991, scattering debris over South America.

THE MIR SUCCESS STORY

The Mir space station is one of the great success stories in space history. Despite the media indulgence in the failures, collisions and fires associated with the space station, its performance has been remarkable during a career which was still continuing in 2000. Mir began life as a single Salyut-type module, and eventually grew into a complex of one small Kvant 1 module plus four more modules, about the same size as the Mir core, called Kvant 2, Kristall, Spektr and Priroda. The Mir space station hosted numerous crews, nine American Space Shuttle missions and was rarely unmanned during its lifetime. The problems that were experienced are to be expected on any space station. The experience gained in handling these problems will be invaluable in the operation of the International Space Station (ISS), which western journalists have already assumed will never have any of the problems experienced by Mir because it is apparently superior and largely US-made.

The 20.9 tonne Mir core module was launched on 26 February 1986, and included a multiple docking port to receive up to five vehicles. A Kvant 1 module, weighing 11 tonnes, docked at the rear of Mir on 9 April 1987. Mir now measured almost 19m (62.3ft) long. The 18.5 tonne Kvant 2 module was added to the front port on 6 December 1989, and was equipped with an array of telescopes, cameras and equipment. Kristall, weighing 19.6 tonnes,

The Mir space station is one of the great success stories in space history. Despite the media indulgence in the failures, collisions and fires associated with the space station, its performance has been remarkable throughout its long career.

165

THE MIR MISSION DIARY

Date	Mission
13 March 1986	Soyuz T15's Leonid Kizim and Vladimir Solovyov make 125 day mission to both Mir and Salyut 7.
5 February 1987	Soyuz TM2 duo includes Yuri Romanenko who makes record-breaking 326-day flight.
22 July 1987	Soyuz TM3's three-man crew includes Syrian cosmonaut during 7-day flight.
21 December 1987	Vladimir Titov and Musa Manarov of Soyuz TM4 make record-breaking 365-day flight.
7 June 1988	Soyuz TM5's three-man crew includes a Bulgarian on 9-day flight.
31 August 1988	Soyuz TM6 carries crew of three including cosmonaut from Afghanistan on 8-day mission featuring an emergency in orbit.
26 November 1988	Two of Soyuz TM7's crew make 151-day flight featuring second flight by Frenchman Jean-Loup Chretien who also makes the first non-US, non-Russian space walk during his 25-day mission.
6 September 1989	Soyuz TM8's two-man crew stays for 166 days.
11 February 1990	Soyuz TM9's two-man crew occupies Mir for 179 days.
1 August 1990	Soyuz TM10's two cosmonauts make 130-day mission.
2 December 1990	Soyuz TM11's main crew of two make 175-day flight. Third cosmonaut, on 7-day mission, is Japanese journalist Toyohiro Akiyama, the first commercial passenger in space.
18 May 1991	Soyuz TM12's commander makes 44-day flight, while flight engineer Sergei Krikalev flies for record 311 days (encompassing the collapse of the Soviet Union and establishment of Russia). Both make record six space walks in 33 days. Crew also includes the UK's Helen Sharman, the first non-Soviet, non-American female in space and the first from Great Britain, during 7-day visit.
2 October 1991	Soyuz TM13 three-man crew includes cosmonauts from Kazakhstan and Austria on 7-day mission.
17 March 1992	Soyuz TM14 makes first official Russian space flight (after collapse of Soviet Union). Crew includes a German on 7-day visit. Main crew remains for 145 days.
27 July 1992	Soyuz TM15's three-man crew includes French cosmonaut who makes 14-day flight.
24 January 1993	Soyuz TM16 two-man crew flies 179-day mission. Alexander Serebrov brings his tally of space walks to a record ten.
1 July 1993	Soyuz TM17 three-man crew includes two who make 196-day mission and a French cosmonaut on 17 day 0hr 45min flight.
8 January 1994	Soyuz TM18's three-man crew includes doctor Valeri Poliakov who stays in space for a record one year and 72 days.
1 July 1994	Soyuz TM19 two-man Russian and Kazakh crew are first all-space rookie crew since 1997 and stay 125 days.
3 October 1994	Soyuz TM20 crew of three includes Russian woman and German whose visit lasts 31 days. Main mission lasts 169 days.
3 February 1995	STS 61 Discovery's six-person crew, including one Russian, makes a rendezvous demonstration flight in preparation for joint Shuttle Mir missions.
14 March 1995	Three man crew aboard Soyuz TM21 includes US astronaut Norman Thagard, the first American to be launched on a Russian rocket. All return on Space Shuttle after 115-day mission.
27 June 1995	Atlantis STS 71 Shuttle Mir Mission (SMM) 1 docks with Mir and crew transfers two Russians launched on the Shuttle. The first joint US/Russian flight.
3 September 1995	Soyuz TM22's three-man crew includes German Thomas Reiter on 179-day flight.
12 November 1995	Atlantis STS 74 flies 8-day SMM 2.
21 February 1996	Soyuz TM23's two-man crew remains for 172 days.
22 March 1996	Atlantis STS 76 SMM 3 delivers Shannon Lucid for 188-day flight.

Date	Mission
17 August 1996	Soyuz TM24's three-person crew includes French female scientist for 16 day flight. Main crew stays 196 days.
16 September 1996	Atlantis STS 79 flies SMM 4 to deliver John Blaha for 128 days and bring back Shannon Lucid.
12 January 1997	Atlantis STS 81 flies SMM 5 delivering Jerry Linenger, who replaces Blaha, for 132 days.
10 February 1997	Crew of three aboard Soyuz TM25 includes German visitor who stays 19 days. 184-day mission hit by accidents including fire and collision.
15 May 1997	Atlantis STS 84 flies SMM 6, delivering Michael Foale who stays 144 days, and brings back Linenger.
5 August 1997	Crew of two aboard Soyuz TM26 are launched on 197 day mission to make urgent repairs.
26 September 1997	Atlantis STS 86 SMM 7 delivers David Wolf, who replaces Foale, for 127 days. First joint US/Russian space walk is made by Vladimir Titov and Scott Parazinski.
23 January 1998	Endeavour STS 89 SMM 8 delivers Andrew Thomas for 141 days, in place of Wolf.
29 January 1998	Soyuz TM27's commander is Kazakh Talgat Musabayev and crew includes a Frenchman who says 20 days. Main crew stays for 207 days.
2 June 1998	Discovery STS 91 is ninth and final SMM. Collects Thomas and includes Valeri Ryumin, the veteran Salyut cosmonaut from Russia and head of Mir programme.
13 August 1998	Soyuz TM28 three-man crew includes Yuri Butarin, a cosmonaut observer and former presidential aide who flies for 11 days. Main mission lasts 198 days. Sergie Avdeyev lands in TM29 after flight lasting a year and 14 days, amassing a record of two years and 17 days in space on three missions.
20 February 1999	The Soyuz TM29 crew includes German cosmonaut on 188 day mission. Slovakian cosmonauts make 7-day flight. Advertised as final Mir mission.
April 2000	New mission to Mir.

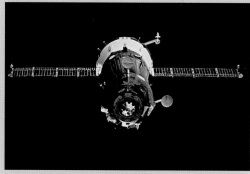

Above: Soyuz TM spacecraft seen from the US Space Shuttle during a fly-around of the Mir space station in 1995.

Left: The Space Shuttle Atlantis is docked to the Mir space station in 1995. Note the size of the Shuttle compared to Mir.

MIR

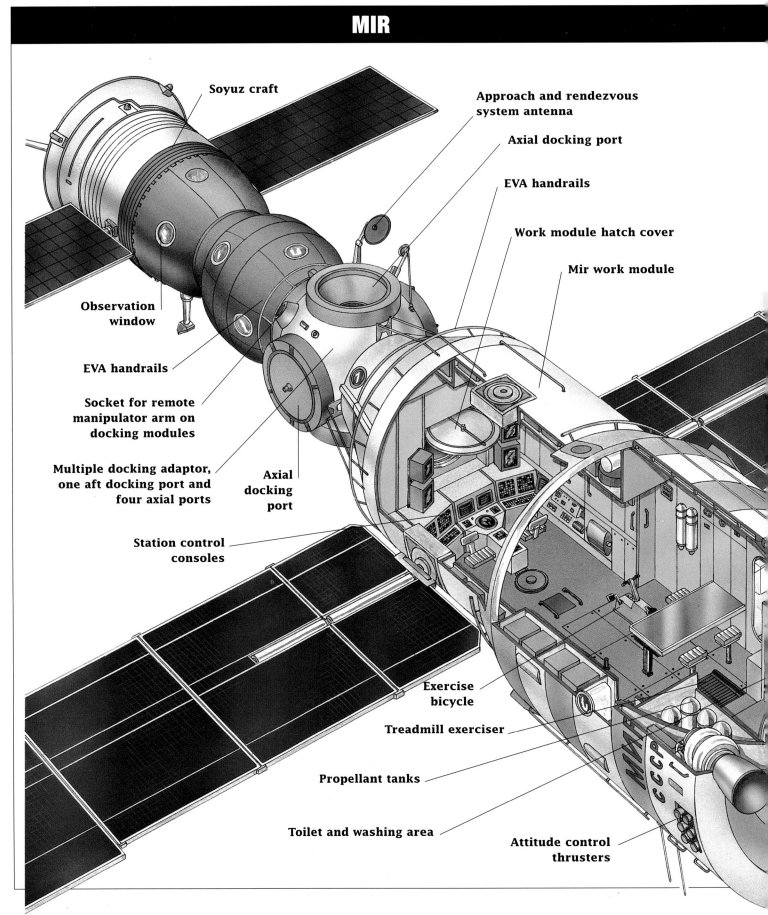

Soyuz craft

Approach and rendezvous system antenna

Axial docking port

EVA handrails

Work module hatch cover

Mir work module

Observation window

EVA handrails

Socket for remote manipulator arm on docking modules

Multiple docking adaptor, one aft docking port and four axial ports

Axial docking port

Station control consoles

Exercise bicycle

Treadmill exerciser

Propellant tanks

Toilet and washing area

Attitude control thrusters

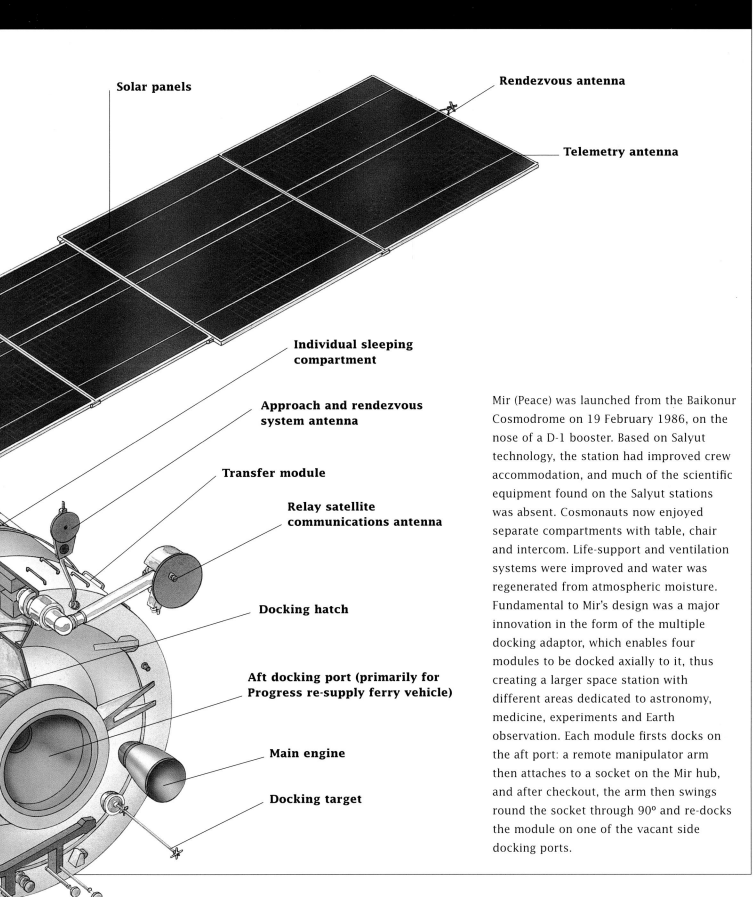

Solar panels

Rendezvous antenna

Telemetry antenna

Individual sleeping compartment

Approach and rendezvous system antenna

Transfer module

Relay satellite communications antenna

Docking hatch

Aft docking port (primarily for Progress re-supply ferry vehicle)

Main engine

Docking target

Mir (Peace) was launched from the Baikonur Cosmodrome on 19 February 1986, on the nose of a D-1 booster. Based on Salyut technology, the station had improved crew accommodation, and much of the scientific equipment found on the Salyut stations was absent. Cosmonauts now enjoyed separate compartments with table, chair and intercom. Life-support and ventilation systems were improved and water was regenerated from atmospheric moisture. Fundamental to Mir's design was a major innovation in the form of the multiple docking adaptor, which enables four modules to be docked axially to it, thus creating a larger space station with different areas dedicated to astronomy, medicine, experiments and Earth observation. Each module firsts docks on the aft port: a remote manipulator arm then attaches to a socket on the Mir hub, and after checkout, the arm then swings round the socket through 90º and re-docks the module on one of the vacant side docking ports.

was next to join the complex on 10 June 1990, being transferred the next day to an adjoining port with the aid of a small robot arm. Kristall was dedicated mainly to experiments processing materials. The next module to join was Spektr on 1 June 1995. Weighing 19.6 tonnes at launch, it was dedicated to Earth sciences and atmospheric monitoring. The final Mir module, the 19.7 tonne Priroda, arrived on 26 April 1996 equipped with an array of remote sensing cameras.

The quite remarkable Mir enabled cosmonauts to record flight times lasting more than a year, and to amass a wealth of biomedical data about the effects of long space flights on the human body. From 1986, a continuous stream of cosmonaut teams visited Mir and were still being sent there in 2000. These teams also included crew members from other countries. Later in the programme, after the collapse of the Soviet Union, Russia began to charge for these foreign trips and experiment time on the station. Regular EVAs were made outside Mir to conduct experiments and make repairs.

In 1995, the first Americans started to visit Mir and live on board. Their presence marked the first opportunity for the US to experience flights of lengthy duration, and to exceed the record of 84 days set by the Skylab 4 crew. This record had already been broken by Soviet cosmonauts and exceeded almost five times over. The presence of US astronauts also focused western media attention on the Mir space station, and its occasional mechanical faults and accidents, thereby giving Mir an undeserved reputation. A number of faults were serious, such as the collision of a cargo tanker which damaged the Spektr module and almost caused a major depressurization. The US missions launched to Mir on the Space Shuttle were effectively rehearsals for the US's joint

Right: This is how the ISS should look when its assembly has reached completion. This is projected for 2004, though almost inevitably there will be delays.

ISS CONFIGURATION

This illustration shows the various nations involved in the construction and assembly of the International Space Station.

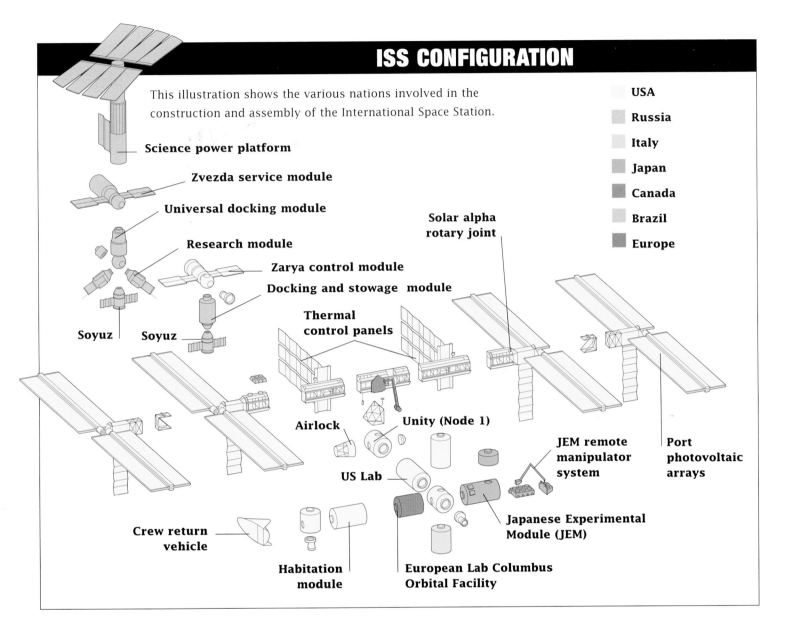

USA
Russia
Italy
Japan
Canada
Brazil
Europe

Science power platform

Zvezda service module

Universal docking module

Research module

Solar alpha
rotary joint

Zarya control module

Docking and stowage module

Soyuz   Soyuz

Thermal
control panels

Airlock

Unity (Node 1)

JEM remote
manipulator
system

Port
photovoltaic
arrays

US Lab

Japanese Experimental
Module (JEM)

Crew return
vehicle

Habitation
module

European Lab Columbus
Orbital Facility

work with the Soviets on the proposed International Space Station (ISS).

THE INTERNATIONAL SPACE STATION

NASA's initial space station hopes had been dashed by budget cuts in the mid-1970s when the space agency had been developing the Space Shuttle. However, the agency pushed hard for an American space station once the Shuttle was operational in the early 1980s, and while the Soviet Union was launching the Mir space station, in a climate not unlike the Cold War. President Reagan had introduced the "Star Wars" Strategic Defense Initiative to counter the threat of Soviet missiles. Reagan gave the

station the go-ahead in 1984, claiming that it would be fully operational by 1994 – provided it was an international affair which involved Canada, Europe and Japan. The budget was set at $5 billion.

Known as Freedom, the station's design was extraordinarily ambitious: a huge dual-keel structure constructed by spacewalking astronauts on dozens of Shuttle flights a year, commencing in 1992. Gradually, it was accepted by NASA and the US Congress that Freedom was too ambitious and could not be built within the budget. It was also already behind schedule. The design was scaled down almost annually to cut the costs, as

INTERNATIONAL SPACE STATION

The ISS is one of the largest international civil co-operative programmes ever attempted, and involves 15 nations – the US, Russia, Canada, Japan, Brazil, Belgium, Denmark, France, Germany, Italy, the Netherlands, Norway, Spain, Sweden, and Switzerland. When completed, the space station will weigh over 453 tonnes and will measure 111.32m (365.2ft) end to end, equivalent in length to a football field. It will be manned by six crew members, and will always include at least one US and one Russian astronaut. A major part of the station will be the Canadian remote manipulator system of two robot arms, one measuring 16.77m (55ft). Also included will be a transporter designed to travel along a rail the length of the station. The ISS will be assembled from modules, nodes, truss segments, arrays of solars, re-supply tugs, and thermal radiators, and will provide 1624m³ (57,351ft³) of pressurized living and working space – the equivalent of the inside of a 747 Boeing jet. The station will have a habitation module, two US laboratories, a European module, a Japanese module, and two Russia research modules, as well as other modules for providing services. Assembly will require 45 rocket launches, principally the Space Shuttle. Four photovoltaic modules, each with two arrays, 34.16m (112ft) long and 11.89m (39ft) wide, will each generate 23kW of electricity. The total surface area of the arrays is about half an acre – 2500m³ (88,286ft³). The electrical power system is connected with 12.81km (8 miles) of wire. The batteries measure 883m (2897ft) lined end-to-end. Connected to the ISS will be cupola modules, each with four windows, offering a 360° view for observing Earth. Fifty-two computers will control ISS systems, including orientation, electrical power-switching and solar-panel alignment.

Above: The Zarya control module (below) and the Unity node module are pictured in 1999 awaiting the attachment of more modules to the end of Zarya and to multiple ports on Unity.

Left: An early spacewalk at the ISS to erect antennae and other equipment on the exterior.

Congress baulked at the billions being spent on a project that was being delayed year on year.

In 1994, when the project was already $25 billion in the red, with nothing in space and no launches in the immediate future, Congress informed NASA that there would be no Freedom unless Russia joined the project. This amazing turn of events came about as a result of the collapse of the Soviet Union. Plans for a Mir 2 had been scrapped due to lack of funds, so both countries needed each other. Both wanted a space station. Russia joined what has now become known as the International Space Station, while NASA's other frustrated partners were effectively sidelined.

Due to continuing problems, the ISS start date was delayed again and again until at last, in October 1998, the first ISS module, the Russian Zarya, was launched. The US Unity Node module was docked to Zarya in December 1998. Another Shuttle flew a maintenance mission in May 1999, but the ISS was left unmanned for almost a year after this as delays continued. The date for the ISS to become fully operational slipped to 2004, already ten years late. A Russian service module, Zvezda, was launched in July 2000 to enable further Shuttle flights to continue the assembly, and to set up the right conditions for the first habitation of an expedition crew. Further delays seem inevitable. The likelihood is that the ISS will be completed but it will not resemble what had been designed and will develop on a step-by-step basis, subject to further delays and the availability of finances.

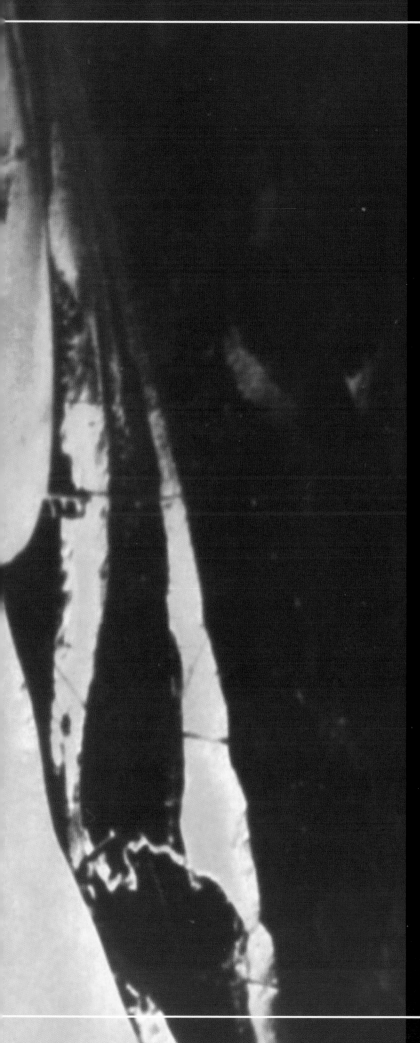

SATELLITES TODAY

Every year dozens of satellites are launched into orbit around the Earth, and other spacecraft are dispatched towards other targets in our solar system, such as Mars, asteroids or comets. Unlike the pioneering years of the Space Age, space flight today has become routine. Few people appreciate that satellites provide vital services which support the day-to-day life of the majority of the population in the industrial world.

Even the developing and less advanced countries of the world are benefiting to some extent from space today. Science satellites such as the Hubble Space Telescope are helping us to appreciate the beauty, vastness and wonders of the universe and the workings of the sun. Communications satellites, such as Galaxy XI, are supporting or providing many kinds of communications services, including international and mobile telephony and, of course, satellite TV, not to mention playing a major role in the rapid growth of the internet. Remote-sensing satellites, such as

Left: The European Eureka multi-purpose science satellite deployed from the Space Shuttle is pictured over the Kennedy Space Centre in Florida.

Spot, are providing services to various industries and organisations, such as helping geologists find areas rich in certain minerals; aiding urban planners in locating new estates; and helping environmentalists to monitor pollution levels in rivers and the sea.

Aircraft, ships and vehicles navigate using Navstar Global Positioning System (GPS) satellites, and road-fleet management is now assisted by data-messaging and positioning satellites. Our TV weather forecasts always feature satellite images from craft such as Meteosat, while other satellites monitor Earth's environment, providing such data as wave height and sea temperature. Military satellites provide the armed forces and security organisations with services which include electronic intelligence monitoring by Magnum satellites, and extremely high resolution images from classified optical and radar spacecraft.

THE HUBBLE SPACE TELESCOPE

An optical telescope on the ground is impeded to some extent by having to peer through Earth's atmosphere – rather like trying to see underwater in a swimming pool. By putting an optical telescope above the atmosphere, its view is greatly enhanced, and its ability to observe more distant objects improved. The Hubble Space Telescope (HST) was taken into orbit by the Space Shuttle Discovery STS 31 on 24 April 1990. HST is in an orbit inclined to the equator by 28.5°, which is almost circular, at about 607km (377 miles) altitude.

The HST was designed to be serviced in orbit by later Space Shuttle crews, and to have instruments removed and replaced, including its two solar arrays, each 12.19m (40ft) long. The 11.6 tonne telescope is 13m (42.7ft) in length and 4.2m (13.7ft) in diameter at its widest. It is equipped with two high-gain antennae, which enable it to transmit data direct to the ground, using a

Tracking and Data Relay Satellite system, and two low-gain antennae. A data-management system, which includes a high-power computer and fine-pointing system, enables the HST to point to and remain locked on any specific target to within 0.01 arc seconds. If HST was located in Los Angeles, it could focus on a dime in San Francisco.

The Optical Telescope Assembly is configured in such a way that the telescope is the equivalent in length to 57.6m (189ft) but is compacted to 6.4m (21ft). Light enters the aperture door and travels down the tube onto a 2.4m (7.8ft) primary mirror, and is then reflected onto a secondary mirror, 0.3m (11.8in) in diameter, before it is reflected through a hole in the centre of the primary mirror onto the focal plane. HST's science instruments then receive the light. HST was originally equipped with a Faint Object Camera (FOC), Wide Field/Planetary Camera (WFPC), Goddard High-Resolution Spectrograph (GHRS), Faint-Object Spectrograph (FOS), High Speed Photometer (HSP) and Fine Guidance Sensors (FGS).

Once the telescope had reached orbit, and despite the fact that some good images of the universe had been taken by the HST, it soon became clear that the primary mirror was not as perfectly curved as intended. The mirror was suffering from a spherical aberration which had occurred during manufacture. HST needed a "pair of glasses". These were duly manufactured and were named the Corrective Optics Space Telescope Axial Replacement unit (COSTAR), which was about the size of a telephone booth. COSTAR was flown to the HST by Space Shuttle STS 61 on 2 December 1993. The unit was fixed inside the telescope during one of a series of five space walks by dual teams of astronauts who also replaced a solar panel, repaired electronics, installed a new computer processor, replaced the WFPC with a new unit, installed magnetometers, removed

Far right: A Space Shuttle spacewalk to service the Hubble Space Telescope. Three Shuttle servicing missions have been flown and another is planned for 2001.

THE HUBBLE SPACE TELESCOPE

The Hubble Space Telescope (HST) was deployed in to orbit by the Space Shuttle in 1990 and has enabled astronomers to peer far deeper into space than with Earthbound telescopes. It consists of three parts: (1) an Optical Telescope Assembly (OTA) containing primary and seconday mirrors; (2) the Scientific Instruments (SI); and (3) the Support System Module which contains a very precise stabilization system and the power system. Electrical power is derived from solar panels. A meteoroid shield and sunshade protect the optics. The open front end of the Space Telescope is similar to most Earth-based telescopes and admits light to the primary mirror in the rear of the telescope. The primary mirror projects the image to a smaller secondary mirror at the front. The beam of light is then reflected back through a hole in the primary mirror to the Scientific Instruments in the rear which provide the means of converting the telescope images to useful scientific data.

SPECIFICATIONS

Length: 13.1m (42.5ft)
Diameter: 4.26m (14ft)
Weight: 11,600kg (25,500lb)
Orbit: 607km (330 miles)

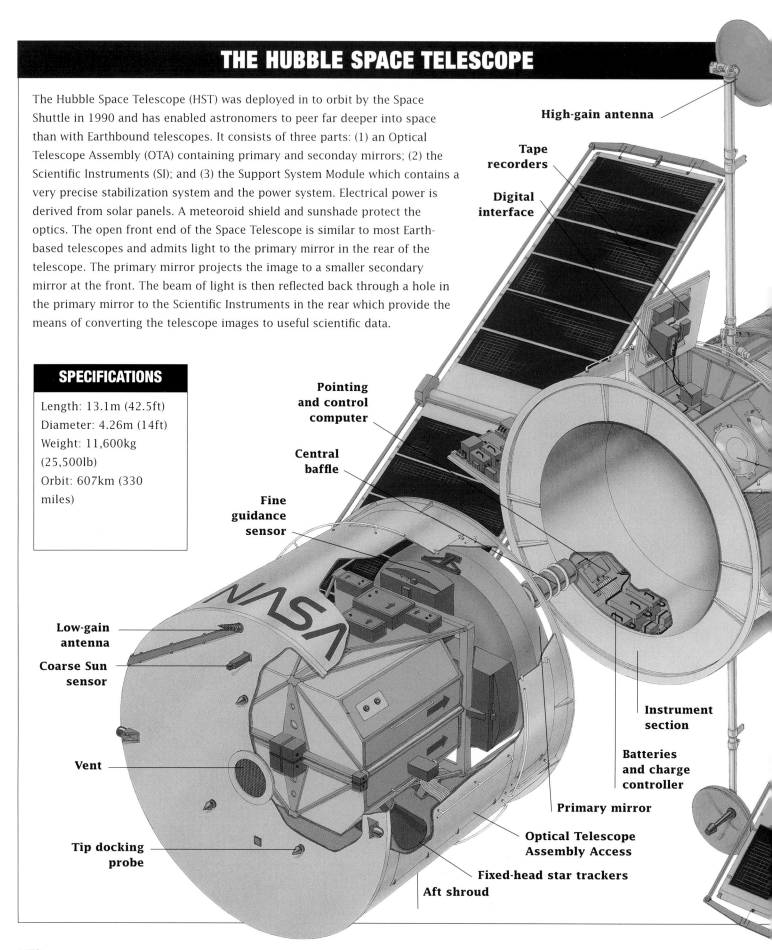

High-gain antenna

Tape recorders

Digital interface

Pointing and control computer

Central baffle

Fine guidance sensor

Low-gain antenna

Coarse Sun sensor

Vent

Tip docking probe

Instrument section

Batteries and charge controller

Primary mirror

Optical Telescope Assembly Access

Fixed-head star trackers

Aft shroud

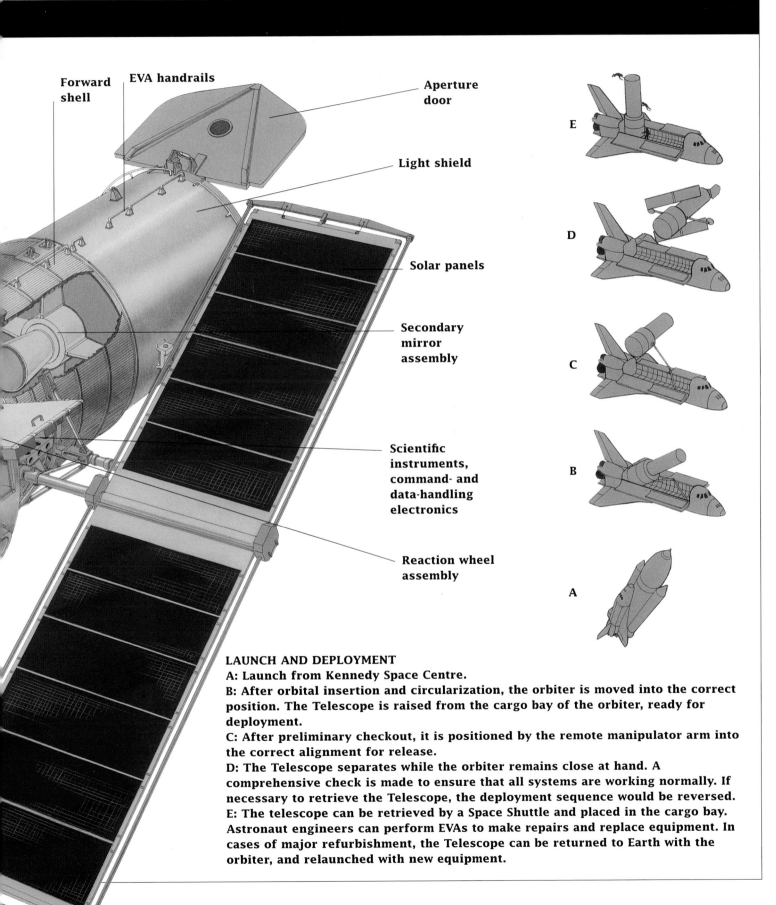

Forward shell

EVA handrails

Aperture door

Light shield

Solar panels

Secondary mirror assembly

Scientific instruments, command- and data-handling electronics

Reaction wheel assembly

E

D

C

B

A

LAUNCH AND DEPLOYMENT
A: Launch from Kennedy Space Centre.
B: After orbital insertion and circularization, the orbiter is moved into the correct position. The Telescope is raised from the cargo bay of the orbiter, ready for deployment.
C: After preliminary checkout, it is positioned by the remote manipulator arm into the correct alignment for release.
D: The Telescope separates while the orbiter remains close at hand. A comprehensive check is made to ensure that all systems are working normally. If necessary to retrieve the Telescope, the deployment sequence would be reversed.
E: The telescope can be retrieved by a Space Shuttle and placed in the cargo bay. Astronaut engineers can perform EVAs to make repairs and replace equipment. In cases of major refurbishment, the Telescope can be returned to Earth with the orbiter, and relaunched with new equipment.

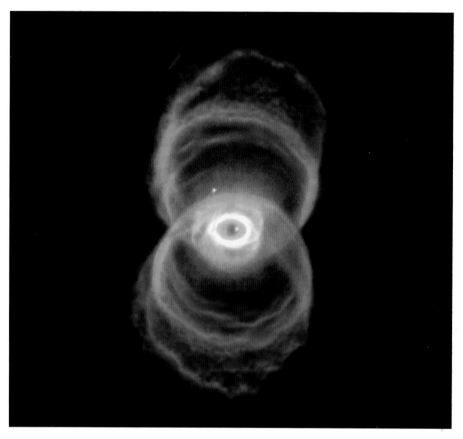

Above: The Hourglass nebula within M8 in Sagittarius, photographed by the HST, showing rings of debris emanating from the remains of an exploded star. The HST has added considerably to our understanding of the universe.

the HSP, and installed a redundancy kit for the GHRS. The results of the installation of COSTAR were immediate. The HST images were now truly spectacular, and the telescope quickly caught the imagination of the media worldwide.

After more than four years of providing astronomers and the public at large with images and data, another planned HST servicing mission by Shuttle astronauts was launched on 11 February 1997. The STS 82 Discovery crew made a comprehensive change to Hubble's suite of instruments, and conducted routine spacewalk servicing. The GHRS and FOS were removed and replaced with a Space Telescope Imaging Spectrograph (STIS) and a combined Near-Infrared Camera and Multi-Object Spectrometer (NICMOS). The astronauts replaced an FGS and other equipment, including tape recorders, installed an optical electronics enhancement kit, and changed the HST stabilisation and fine-pointing reaction-

wheel assemblies. Other work involved laying new thermal insulation blankets.

HST was working as it entered its 10th year of service when it was time to launch a new Shuttle servicing mission. This mission had originally been planned for 2000 but when gyroscopes on the telescope failed to a critical level, the mission was brought forward to December 1999 and split into two, with the second half of the mission put back to 2001. STS 103 Discovery was launched on 19 December 1999, and the mission installed six new gyros and voltage/temperature kits, a new computer 20 times faster and with six times the memory, a new digital tape recorder, and replaced one FGS and a radio transmitter. Over a series of EVAs astronauts also placed new insulation over part of the HST. Two further Shuttle servicing missions are planned to install new solar arrays as well as an Advanced Camera for Surveys, a Cosmic Origins Spectrograph and a third WFPC.

HIGH-POWER COMMUNICATIONS

Geostationary orbit (GEO) is one of the busiest places in space, with hundreds of communications satellites spread around the globe, providing people with telephone, fax, data, TV, messaging and other services. Typical of this sector of the communications-satellite business is a satellite called Galaxy XI, launched on an Ariane 4 booster in 1999. Galaxy resembles a box with many kinds of aerials, dishes and other equipment mounted on it. The satellite is located at 91°W over the equator and serves customers in North America and Brazil. It is operated by an international communications provider called PanAmSat, which is currently the world's leading commercial provider of satellite-based communications services, operating a global network of 20 satellites.

Galaxy XI is equipped with transponders which operate in designated frequency wavebands. Transponders are combined

COMSTAR 1

One of the first geostationary satellites in a domestic role was Comstar. Operated by COMSAT General and leased by American Telephone and Telegraph Company, the Comstar 1 domestic satellites were designed to operate for seven years. They received, amplified and retransmitted telephone calls and TV broadcasts between ground stations within the US, as well as Puerto Rico. They could relay up to 6000 telephone calls of 12 TV programmes, or combinations of the two.

Bearing and power transfer assembly. A bearing is fitted betweeen the spinning tub and the top section which does not spin because the antennae must be kept pointing at the Earth

Drum spins at about 60rpm to give gyroscopic stability; surface is covered with solar cells which generate electricity from sunlight

Sun and Earth sensors, the reference devices by which Comstar is kept in position

Axial jets

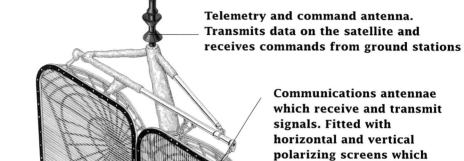

Telemetry and command antenna. Transmits data on the satellite and receives commands from ground stations

Communications antennae which receive and transmit signals. Fitted with horizontal and vertical polarizing screens which allow the same frequency to be used twice, thus doubling the effective capacity of the satellite

Electronic equipment compartment with communications receivers, amplifiers and transmitters

Positioning and orientation system

Battery pack which stores electricity from solar cells to power the satellite when in Earth's shadow

Booster adapter

Apogee motor, which lifts satellite into geostationary orbit after separation from launch vehicle

SPECIFICATIONS

Height: 5.2m (17ft)
Diameter: 2.3m (7.5ft)
Launch weight: 1410kg (3109lb)

A YEAR OF SATELLITE AND SPACECRAFT LAUNCHES

Date (1999)	Launches
3 January	US Mars Polar Lander
27 January	US/South Korea Rocsat 1 communications satellite
7 February	US Stardust solar system explorer
9 February	Four Globalstar international mobile-phone communications satellites
16 February	Japanese JCSAT 6 geostationary orbit (GEO) communications satellite
20 February	Russian Soyuz TM29 Mir ferry
23 February	US Argos military technology satellite
26 February	UK Skynet 4E military communications satellite Saudi Arabian Arabsat 3A GEO communications satellite
28 February	Russian Raduga 1 GEO communications satellite
5 March	US Wire infrared astronomical explorer satellite
15 March	Four Globalstar international mobile-phone communications satellites
21 March	Hong Kong Asiasat 3S communications satellite
28 March	Demonstration satellite for new Sea Launch boosters
2 April	Russian Progress M41 Mir tanker Insat 2E Indian GEO communications and weather satellite
10 April	US DSP 19 military early warning satellite
12 April	European Eutelsat W3 GEO communications satellite
15 April	Four Globalstar international mobile phone communications satellites US Landsat 7 remote-sensing Earth observation satellite
21 April	UK UoSAT 12 mini-satellite technological demonstration satellite
29 April	Megsat Italian technology satellite German Axibras X-ray astronomy satellite
30 April	Milstar 2 US military communications satellite
5 May	Orion 3 US commercial communications satellite
10 May	Chinese Fengyun 1C meteorological satellite
17 May	Terriers US students' ionosphere research satellite Mulbcom US military communication technology satellite
20 May	Nimiq 1 Canadian GEO communications satellite
22 May	US National Reconnaissance Office satellite
26 May	Oceansat 1 Indian Ocean observation satellite
27 May	Discovery STS 96 ISS mission
10 June	Four Globalstar international mobile-phone communications satellites
11 June	Two Iridium international mobile-phone communications satellites
18 June	Astra 1H Luxembourg-based direct TV GEO communication satellite
20 June	Quicksat US remote-sensing technology satellite
24 June	Fuse US ultraviolet astronomy satellite
8 July	Russian Molniya 3 communications satellite

Above: A 1m resolution close-up of the Coliseum in Rome, Italy, seen from orbit by the Ikonos commercial remote sensing satellite, equivalent to what a spy satellite can discern.

Date	Launches
10 July	Four Globalstar international mobile-phone communications satellites
16 July	Russian Progress M42 Mir tanker
17 July	Okean 0-1 Russian ocean observation and monitoring satellite
23 July	Columbia STS 93 deployment of Chandra X-ray observatory
25 July	Four Globalstar international mobile-phone communications satellites
12 August	Telkom 1 Indonesian GEO communications satellite
17 August	Four Globalstar international mobile-phone communications satellites
18 August	Russian Cosmos 2365 military reconnaissance satellite
26 August	Russian Cosmos 2366 navigation satellite
4 September	Koreasat 3 South Korean GEO communications satellite
6 September	Two Yamal Russian GEO communications satellites
9 September	Foton 12 international microgravity recoverable research satellite
22 September	Four Globalstar international mobile-phone communications satellites
23 September	US Echostar V GEO communications satellite
24 September	US Ikonos commercial high-resolution remote-sensing satellite
	US Telstar 7 GEO communications satellite
26 September	LM-1 US/Russian GREO communications satellite
28 September	Resurs F1M Russian remote-sensing satellite
7 October	US Global Positioning System (GPS) Block IIR navigation satellite
9 October	US DirecTv 1R GEO communications satellite
14 October	Chinese/Brazilian CBERS remote-sensing satellite
18 October	Four Globalstar international mobile-phone communications satellites
19 October	US Orion 2 GEO communications satellite
13 November	US GE-4 GEO communications satellite
20 November	Chinese unmanned test of the Shenzou manned spacecraft
22 November	Four Globalstar international mobile-phone communications satellites
22 November	US Navy UHF GEO communications satellite
3 December	French spy satellite Helios 1B
4 December	Five Orbcomm data-messaging satellites
10 December	Europe's XMM X-ray astronomical telescope
12 December	US DSMP 5D military meteorological satellite
18 December	US Terra polar-orbiting Earth observation platform
19 December	Discovery STS 103 Hubble Space Telescope servicing mission
21 December	Kompsat South Korean multi-purpose satellite
22 December	US Galaxy XI GEO communications satellite
26 December	Russian Cosmos 2367 ocean electronic intelligence satellite
27 December	Cosmos 2368 missile launch early warning satellite

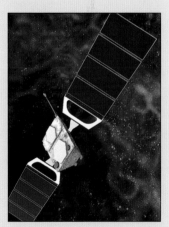

Above: A Globalstar worldwide mobile communications satellite system spacecraft. The fully operational Globalstar satellite constellation numbers 48 spacecraft in low Earth orbit.

SOVIET SATELLITE SYSTEMS

The Soviet Union was first to use satellites domestically with their Molniya satellites in 12-hour orbits which allowed the relay of a large volume of radio, TV, telephone, telex and fax traffic. Ekran ("Film") was introduced in October 1976 to transmit TV programmes to isolated communities in Siberia and the Far North. The Soviet Union used two orbital systems for internal and external communications. One was the highly eccentric orbit approx 40,000km x 500km (24,856 miles x 311 miles) inclined at about 65° to the Equator in which active Molniya satellites with approximately equal spacing provided 24-hour service in the Northern Hemisphere. The other was geostationary orbit used by Ekran.

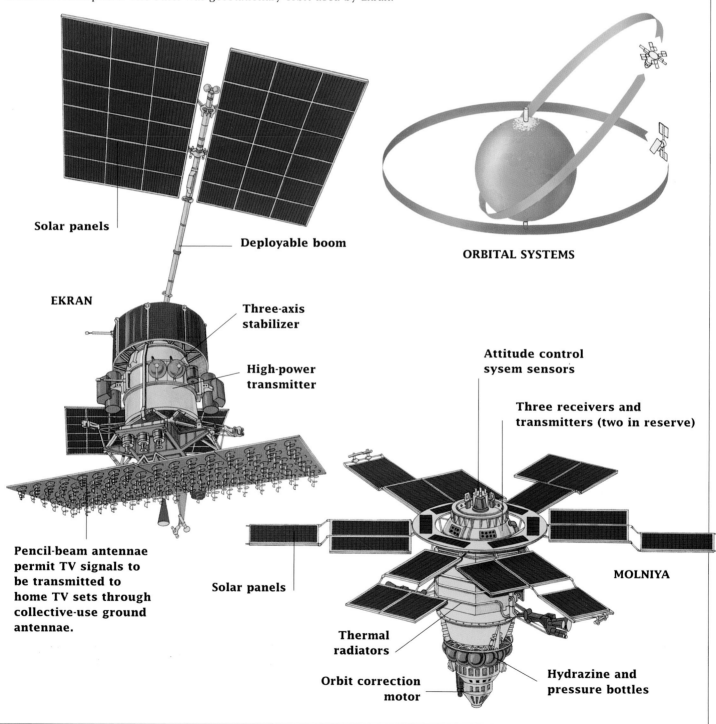

Solar panels

Deployable boom

ORBITAL SYSTEMS

EKRAN

Three-axis stabilizer

High-power transmitter

Attitude control sysem sensors

Three receivers and transmitters (two in reserve)

Pencil-beam antennae permit TV signals to be transmitted to home TV sets through collective-use ground antennae.

Solar panels

MOLNIYA

Thermal radiators

Orbit correction motor

Hydrazine and pressure bottles

Left: A close-up of the solar array on a typical communications satellite. Thousands of solar cells on the arrays generate electrical power for the energy-consuming payloads.

transmitters and receivers which receive radio signals and automatically re-transmit them, often at a different frequency. The signals transmitted to the satellite are weak by the time they reach the satellite, and have to be amplified by travelling wave tubes. The transmissions are received by antennae and sent back to Earth using the exquisitely designed dish aerials or reflectors which direct the signals to specific areas on Earth, called footprints.

The satellite is equipped with 20 C-band

GEOSTATIONARY ORBIT

A satellite in geostationary orbit (GEO) matches the period of the Earth's rotation above the equator. It moves neither east nor west relative to a fixed point on the surface. A single satellite, therefore, can provide uninterrupted communications service 24 hours a day. Lower altitude systems, like the Telstars, required that ground stations should follow the satellites as they crossed the sky. Then the stations had to switch from satellite to satellite as one slipped below the horizon and another came within range. The first steps toward GEO were taken by NASA with Syncom 2, an inclined satellite which traced a figure-of-eight ground track, and Syncom 3, a geostationary satellite.

Telstar 1

Telstar 2

Syncom 3 in GEO

Syncom 2

Right: The Galaxy XI
communications satellite in
geostationary orbit over its
service area provides a
multitude of services to
users in the Americas. A
satellite in GEO is facing
one third of the Earth's
hemisphere.

and 40 Ku-band transponders. The C-band transponders are used to serve cable-TV customers, while the Ku-band payload is used for video distribution, data networks and other general communications services. To operate a satellite with so many transponders requires an enormous amount of electricity, which is generated

by huge arrays of solar cells stretched like wings – with a wingspan of 62m (203ft), longer than the wings of the Boeing 747 – from both sides of the satellite. These cells are made of gallium arsenide, an alternative to traditional silicon cells, which provide twice the power capability. The panels are folded against the spacecraft for launch and then deployed in orbit. They feature angled reflector panels along both sides of the arrays which form a shallow trough to concentrate the Sun's rays on the solar cells. Galaxy XI can operate using 10kW of generated power, and is the largest and most powerful civilian commercial communications satellite yet launched. It weighs 2.77 tonnes in space.

Meanwhile, a constellation of low Earth orbiting (LEO) satellites is helping to support the provision of worldwide mobile phone communications, including voice-messaging, fax and data transmission. A Globalstar satellite system provides satellite telephone services for cellular phone users who roam outside coverage areas; for people who work in remote areas where terrestrial systems do not exist; for residents of under-served markets, usually in developing nations, who can use fixed-site phones to satisfy their needs for basic telephony; and for international travellers who need to keep in constant touch.

The phones look and act like mobile or fixed phones, but the difference is that they can operate virtually anywhere, carrying a call over an exceptionally clear satellite signal. Like "bent-pipes", or mirrors in the sky, the Globalstar constellation of 48 LEO satellites pick up signals from over 80 per

Above: A Tempo communications satellite is equipped with dozens of transponders and dish aerials, providing multiple customers with a range of services, from direct-to-home TV pictures to multimedia services.

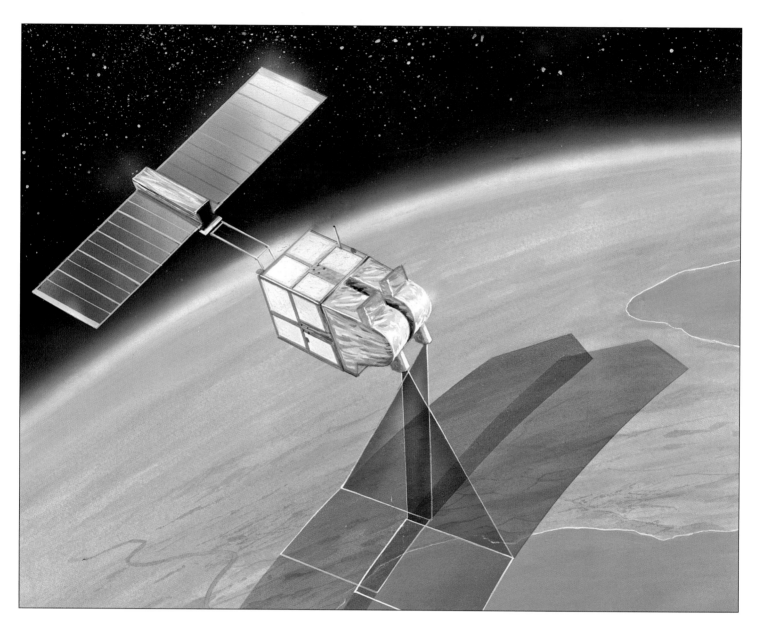

Above: A French Spot remote sensing satellite takes a series of multispectral images of the Earth in swaths across the planet's surface.

cent of the Earth's surface, everywhere except extreme polar regions and some mid-ocean areas. Several satellites pick up a call, and this "path diversity" assures that the call is not lost even if a phone moves out of sight of one of the satellites.

ON-THE-SPOT EARTH OBSERVATION

One of the main uses of satellites is to provide images of Earth. Taken high above Earth from space, these images show a large area in great detail. The satellites are called remote-sensing satellites, because remote sensing is the collection of information or mapping of a certain area

remotely rather than being physically on the spot. Satellites in polar orbits can cover the whole Earth in a single day, so they can save time and money compared with systems that take much longer to cover the same area, such as aircraft operating at a lower altitude. Some satellite instruments are so powerful that from a height of 480km (298 miles), they can discern objects on the ground as small as 1m (3.3ft) across. Satellites carry an array of sensitive cameras which can take many types of "multispectral" images, such as those that can monitor the temperature on parts of the Earth's surface or types of vegetation.

LANDSAT 4

At the time of its launch on 16 July 1982, Landsat 4 was the most advanced of NASA's Earth resources satellites. The satellite considerably improved the technology of remote sensing of the Earth, thereby aiding resource management. Landsat 4 was the first NASA satellite to incorporate a Global Positioning System (GPS). Using data supplied by navigation satellites, Landsat's computer could calculate the craft's position and velocity, signals being received by means of a GPS antenna.

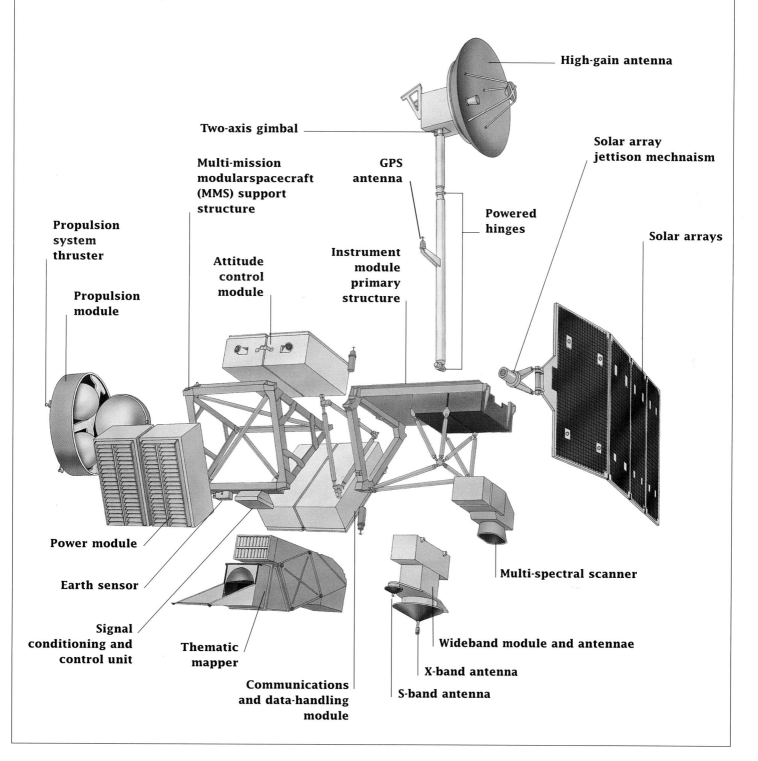

Above: A technician prepares a Global Positioning System Navstar satellite for launch. Navstars can aid users to determine their position on the Earth's surface to within 30m (98.4ft) or less.

Some of these instruments, such as radar and infrared, can even penetrate darkness and "see" at night. Other satellites can peer 9m (29.5ft) under water.

A picture taken from a satellite is not of much use on its own. The satellite remote-sensing industry is not so much about what an image can do but what can be done with the image in order to make it useful. Reproducing information and maps from raw satellite images is made possible by the use of powerful computer technology. Remote-sensing data is transformed into products, by being manipulated and enhanced by computer. The maps and images produced as a result combine data taken from several types of image of the same area and data from other sources, which may include ground surveys and existing maps. Through this work, valuable data products are provided to many different industries including oil exploration and cartography, environmental monitoring and urban planning.

One of the commercial companies that operates a fleet of remote-sensing satellites is called Spot Image, a mainly French concern which claims 60 per cent of the commercial market for images. Four Spot satellites have been launched to date on Ariane boosters, the latest in 1998. Spot 4 weighs 2.75 tonnes and is equipped with two solar arrays which generate 2.1kW of electrical power for the instruments. These comprise two high-resolution imagers which operate in two wavebands of the electromagnetic spectrum of radiation, visible light and infrared, and offer a resolution of 10m (32.8ft) at best, each image covering a swath measuring 60km² (23 square miles).

A vegetation instrument operating in the short-wave infrared band can return 1km (0.6 mile) resolution images covering a swath of 2200km (1367 miles). This is used largely by the European Union to monitor crops, and to ensure farmers are obeying the quota regulations. Spot 4 also carries other instruments, including one to measure polar ozone and aerosol, and a laser communications system to demonstrate the ability of one satellite to communicate with another in orbit. The satellite is programmed to cover certain areas of Earth on request, and transmits its stored images to customized ground stations all over the world.

GLOBAL POSITIONING SYSTEM (GPS) AS A WAY OF LIFE

A fleet of at least 24 US-built and operated Navstar Block IIA and IIR GPS satellites have been launched on Delta boosters, and are always in orbit in constellations consisting of six orbital planes. Each plane has a minimum of four satellites, placed in circular orbits of approximately 20,000km (12,427 miles), inclined to the equator at 54°, and making one orbit of Earth every 12 hours. The satellites are spaced in orbit so that at any time a minimum of six satellites will be in view to users anywhere in the world.

Emitting continuous navigation signals in two L band frequencies, these satellites are available to all types of military and

SEASAT 4

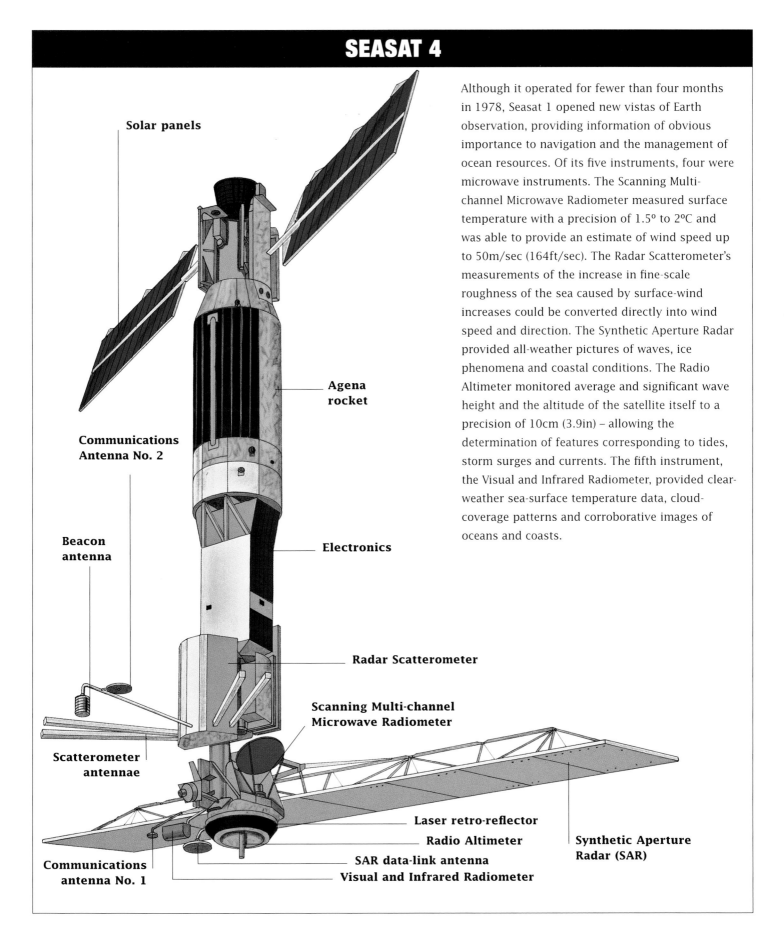

Solar panels

Agena rocket

Communications Antenna No. 2

Beacon antenna

Electronics

Radar Scatterometer

Scanning Multi-channel Microwave Radiometer

Scatterometer antennae

Laser retro-reflector

Radio Altimeter

Communications antenna No. 1

SAR data-link antenna

Visual and Infrared Radiometer

Synthetic Aperture Radar (SAR)

Although it operated for fewer than four months in 1978, Seasat 1 opened new vistas of Earth observation, providing information of obvious importance to navigation and the management of ocean resources. Of its five instruments, four were microwave instruments. The Scanning Multi-channel Microwave Radiometer measured surface temperature with a precision of 1.5° to 2°C and was able to provide an estimate of wind speed up to 50m/sec (164ft/sec). The Radar Scatterometer's measurements of the increase in fine-scale roughness of the sea caused by surface-wind increases could be converted directly into wind speed and direction. The Synthetic Aperture Radar provided all-weather pictures of waves, ice phenomena and coastal conditions. The Radio Altimeter monitored average and significant wave height and the altitude of the satellite itself to a precision of 10cm (3.9in) – allowing the determination of features corresponding to tides, storm surges and currents. The fifth instrument, the Visual and Infrared Radiometer, provided clear-weather sea-surface temperature data, cloud-coverage patterns and corroborative images of oceans and coasts.

DYNAMIC SYSTEMS

Just how highly detailed satellite observations of Earth will be in the future is illustrated by a new NASA programme that will study some of Earth's more dynamic atmospheric phenomena. One of the satellites, called the Volcanic Ash Mission, or Volcam, is a pathfinder mission for demonstrating the operational and scientific applications of monitoring volcanic clouds and aerosols from a geostationary orbit.

Volcanic clouds are a potential hazard to jet aircraft. Several instances of damage to commercial airliners by volcanic ash have occurred, and in at least one case almost led to a crash. Another craft, called Picasso, will monitor clouds and small atmospheric particles, known as aerosols, as well as their impact on Earth's radiation "budget" – the balance of solar energy which reaches Earth and is then lost to space, a process which ultimately controls the temperature of our planet. It will employ innovative Light-Detection And Ranging (LIDAR) instrumentation to profile the vertical distribution of clouds and aerosols, while another instrument will simultaneously image the infrared (heat) emission of the atmosphere. During the daylight half of its orbit, Picasso will measure the reflected sunlight in an oxygen-absorption band and take images of the atmosphere with a wide-field camera. A CloudSat spacecraft will study the effect of thick clouds on the reduction of Earth's radiation budget. It will use advanced cloud-profiling radar to provide information on the vertical structure of highly dynamic tropical cloud systems. This new radar will enable measurements of cloud properties for the first time on a global basis, evolving our understanding of cloud-related issues.

Above: An image of part of the British Isles taken from a satellite, showing the temperature of the sea. Note the warm "orange" Gulf Stream on the left, and the cold "blue" North Sea.

civilian users worldwide who, with the appropriate equipment, can receive the signals to calculate time, velocity and location to an accuracy of a millionth of a second, a fraction of a kilometre per hour, and a location to within about 30m (98.4ft). GPS receiver units are onboard aircraft, ships, land vehicles, on man-packs and in hand-held units.

The GPS service is used to support land, sea, and airborne navigation, surveying, geophysical exploration, mapping and geodesy, vehicle location systems, aerial refuelling and rendezvous, search and rescue operations. In the civilian sector new applications are continually emerging in addition to existing usage by commercial airlines, tracking fleets, law enforcement agencies, fishermen, hikers and even tractor drivers on farms.

The concept of GPS was based upon satellite ranging, pioneered by the first Transit satellites 40 years ago. Each GPS satellite transmits an accurate position and time signal, and the user's receiver measures the time delay for the signal to reach the receiver, which is the measure of the apparent range of the satellite. Measurements collected simultaneously from at least four satellites in line with the user are processed to provide the three dimensions of position, velocity and time. The receivers then display the user's position, velocity and time, and some additional data, such as distance and bearing to selected waypoints or digital charts. The more recent Navstar Block 2R satellites weigh more than a tonne in orbit, and are based on a box-like spacecraft bus which measures 1.52m (4.9ft) by 1.93m (6.33ft) by 1.91m (6.27), with a twin solar array spanning 19.3m (63.32ft), and generating 1.1kW.

The GPS system is augmented by the former Soviet Union's Global Navigation Satellite System (GLONASS) fleet. Ideally the GLONASS constellation is composed of 24 satellites, eight in each of three orbital

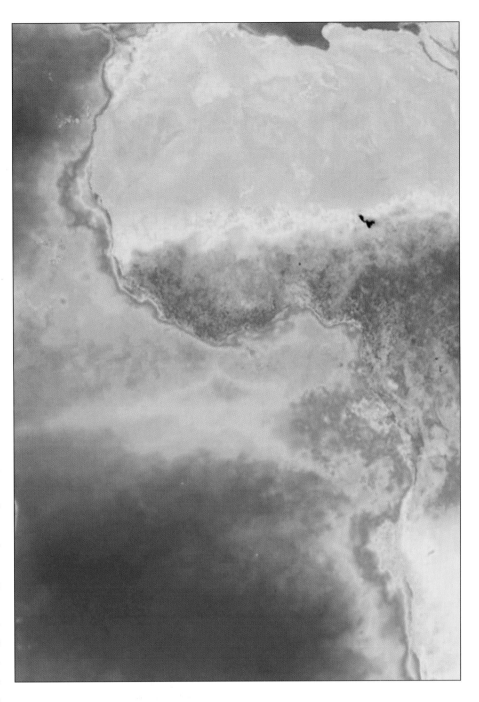

planes, operating in circular orbits of 19,100km (11,868 miles), inclined at 64° to the equator, making one orbit every 11 hours 15 minutes. Like the GPS system, signals can be used by civil and military users, and can be encrypted.

Above: Part of a worldwide map of the vegetation cover of Earth.

THE WORLD WEATHER WATCH

The leading space nations of the world operate a co-operative fleet of polar orbiting and GEO satellites which cover the

NAVSTAR

Each Navstar user set includes a radio receiver with an omni-directional antenna, a signal processor and a readout unit. As the sets operate passively, an unlimited number of users can engage the system without saturating it or revealing their position. The set will automatically select four satellites most favourably located, lock onto their navigational signals and compute the approximate range of each. It will then form four simultaneous equations with four unknowns (the three coordinates of the user's position and the clock bias factor). A small computer in the set will solve the equation for the user's actual position and the time, and determine his velocity. This is what is known as the Global Positioning System (GPS).

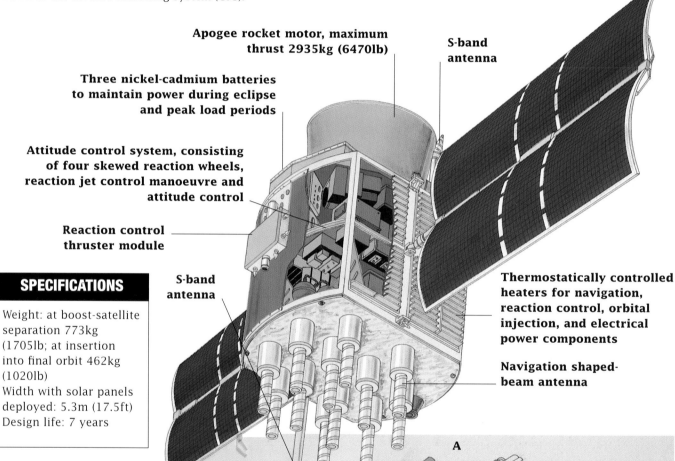

Apogee rocket motor, maximum thrust 2935kg (6470lb)

S-band antenna

Three nickel-cadmium batteries to maintain power during eclipse and peak load periods

Attitude control system, consisting of four skewed reaction wheels, reaction jet control manoeuvre and attitude control

Reaction control thruster module

S-band antenna

Thermostatically controlled heaters for navigation, reaction control, orbital injection, and electrical power components

Navigation shaped-beam antenna

SPECIFICATIONS

Weight: at boost-satellite separation 773kg (1705lb; at insertion into final orbit 462kg (1020lb)
Width with solar panels deployed: 5.3m (17.5ft)
Design life: 7 years

NAVSTAR SYSTEM
This is composed of three integrated segments.
A: The space segment which transmits very accurate satellite position coordinates and timing information.
B: The user segment which processes the time and position data from four satellites.
C: The control segment which will track all the satellites and daily correct their position coordinates and rubidium atomic clocks. The atomic clocks lose or gain an average of only one second in 30,000 years.

A

B

Monitor stations

Master control stations

C

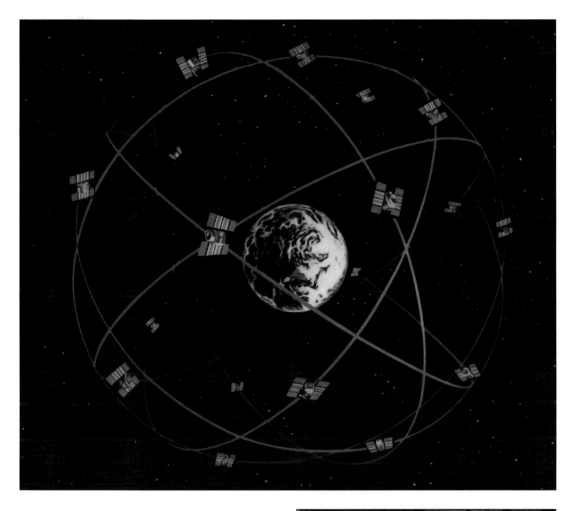

whole globe, returning meteorological and environmental data 24 hours a day. Data ranges from the visible wavelength images of earth that we sometimes see on TV to surface temperature maps. This World Weather Watch system is provided by US, Russian, Chinese, Indian, and Japanese satellites, as well as a fleet of meteorological satellite (Meteosat) GEO spacecraft operated by the European meteorological satellite organisation, Eumetsat, in conjunction with the European Space Agency (ESA). Eumetsat, which operates a Meteosat satellite system, is also planning to launch a fleet of polar orbiting spacecraft, called Metop. The first Meteosat was launched by ESA in 1977, and has been followed by many more. Eumetsat is developing a Meteosat Second Generation (MSG) fleet. These are based on the drum-shaped, spin-stabilised first generation fleet but will be larger, weighing 1.75 tonnes and

Above: All the Navstar satellites circle in polar orbit in one of six planes 20,000km (12,500 miles) above the Earth. They beam signals continuously which allow anyone with a receiver – on land, sea, or in the air – to find his position within 10ft (3m).

Left: A European Meteosat is prepared for launch. A fleet of Meteosat spacecraft take multispectral images of the Earth to monitor phenomena ranging from temperature to moisture content.

ELUSIVE IMAGE

The first image of Earth's entire magnetosphere will be obtained by the Imager for Magnetopause-to-Aurora Global Exploration (IMAGE) satellite, launched in 2000. The magnetosphere is formed by the interaction of the solar wind, ionised or charged plasma particles, mainly of protons and electrons, which stream out at colossal speeds from the sun within the Earth's magnetic field. The magnetosphere is an immense cloud of ionized gas like an invisible fog. Interference in the magnetosphere disrupts terrestrial communications and can result in massive power outages on satellites, costing millions of dollars. The behaviour and extent of the magnetosphere is influenced by changes in the intensity and content of the solar wind. The solar-terrestrial effect will be better understood if scientists can watch while the magnetosphere changes. This will improve predictions of interference on Earth.

The solar wind moves at speeds of up to 900km per second (559 miles per second), and when it encounters earth's magnetic field it is greatly distorted and produces the magnetosphere. It begins about 60,000km (37,500 miles) away from the sunward side of Earth. There is a shock wave or bowshock where the solar wind first meets the magnetic field. Inside the bowshock is a turbulent region of ionization generally referred to as the magnetosheath. On the opposite side of the Earth, the magnetosphere is drawn out to a great distance in the wake of the solar wind.

Activity in the magnetosphere results in the formation of the aurora borealis, more popularly known as the Northern Lights. The phenomena in the southern

hemisphere is known as the Aurora Australis. Electrons and ions travel rapidly back and forth along the lines of Earth's magnetic field. When they are energised by a severe solar wind storm, these particles can punch through and hit the outer atmosphere, thereby creating the spectacular aurorae.

Above: IMAGE is prepared for launch. The satellite will make an intensive study of the Earth's magnetosphere. Interference in the magnetosphere can disrupt communications and result in severe power outages.

the first will be launched by an Ariane 5 in 2001.

The MSGs will be equipped with two major instruments, including an enhanced version of the visible and infrared imager on the earlier Meteosats. The imager scans Earth every 15 minutes in 12 wavelength channels, including a 1km (0.62 mile) resolution for the visible channel and 3km (1.86 mile) for the 11 infrared channels which reveal other atmospheric characteristics, including water-vapour content and ozone levels. The second instrument is the Global Earth Radiation Budget unit which will measure water vapour and cloud-forcing feedback, which are two of the most important processes in predicting climate.

These instruments on the MSGs illustrate that these satellites are more than just craft which send weather images for TV forecasts. The US space agency NASA has already launched the first satellite in a series of Earth Observing System (EOS) craft in its "Mission to Planet Earth" programme. Called Terra, it is a huge polar-orbiting platform weighing 5.19 tonnes, and is 5.9m (19.36ft) long. This and other EOS satellites carry a whole range of instruments which will measure various parameters which affect Earth's weather and climate. These include measurements of the Sun's radiation and its effect on climate, and the hydrological-cycle parameters of atmospheric water content, rain rate, soil moisture, ice, snow, sea and surface temperatures. Other instruments will measure aerosols in the atmosphere, ice sheet topography, ozone depletion, and chlorofluorocarbons, carbon monoxide and methane distribution in the atmosphere. Earth observation is the space sector represented by more satellites than any other.

SPIES IN SPACE

Military operations such as the conflict in Bosnia in 1999 are supported by a fleet of satellites providing services ranging from

communications to missile early-warning systems. Fleets of different kinds of communications satellites enable naval ships, aircraft and ground troops to communicate with headquarters – a troop commander in the field could talk directly to the President in the White House – and to receive the latest intelligence data from

Above: NASA's first Earth Observation System polar platform, Terra, was launched in 1999 to inaugurate an intensive programme of monitoring of the Earth and its atmosphere.

197

Below: A Defence Support System (DSP) early warning satellite is equipped with infra-red sensors to detect the heat from the emission of rocket exhaust or even the heat from the afterburner of a jet fighter.

other satellites, such as high resolution images from spy satellites. These spacecraft take digital images, relayed directly – or via data relay satellites – to the National Reconnaissance Office in Washington and other locations. Some take high-resolution radar images, which can be taken day or night, and even when cloud obscures a target. Data-messaging satellites enable intelligence gatherers to receive messages and transmit them to their headquarters. Troops, ships, aircraft and missiles are guided to their targets by navigation satellites, such as Navstar. Ocean reconnaissance satellites monitor fleet movements. Other satellites are used to test technologies required for the development of a missile defence system, such as those with sensors used to track missiles.

Electronic intelligence satellites, called "elints", such as the CIA's Magnum, monitor radio transmissions and radar emissions from military installations – and can even be used to bug civilian telephone calls. Specific targets include microwave and other communications from military installations, missile telemetry trans-mitted during tests, communications intelligence and radar installations. For example, elints record radio and radar transmissions from military

Below: A Defence Support System (DSP) early warning satellite is equipped with infra-red sensors to detect the heat from the emission of rocket exhaust or even the heat from the afterburner of a jet fighter.

Left: An image of Farnborough in the UK, taken by a Soviet spy satellite, revealing objects as small as cars on a runway. The air base is the venue for the Farnborough International Air Show.

areas. When transmitted to ground stations, the data is replayed and the radar signatures, such as pulse reception, frequency, pulse width, transmitter frequency and modulation, enable the most likely function and method of an installation operation to be identified.

Some elints are like giant "vacuum cleaners" which collect transmissions from many sources through the use of a huge dish receiver. The data can then be unscrambled in ground centres, such as the GCHQ in Cheltenham in Gloucester. Three Magnum/Orion satellites were deployed on Space Shuttle missions and by Titan IV Centaur boosters. They each employed a receiver dish up to 100m (328ft) in diameter.

SPACE EXPLORATION

The planets were once mysterious, unknown worlds – until the Space Age. Mercury, the innermost planet in the solar system, has been explored by just one spacecraft three times. America's Mariner 10 was launched in November 1973, and made one fly-by of Venus en route to a course in orbit around the Sun. Its journey took it close to Mercury in March and September of 1974 and in March 1975, making it the only spacecraft to perform such a feat.

Mariner 10 weighed 503kg (1109lb) and had an octagonal-shaped bus 4.6m (15.1ft) high and 1.38m (4.5ft) in diameter. This held a fuel tank, thrusters for positional control, a sun shield, spectrometers and other instruments, and, like the eyes of an alien, two TV cameras which generated 700-line images transmitted to Earth by a 1.19m (3.9ft) diameter antenna. The spacecraft was also equipped with two magnetometers on a 6m (19.69ft) long boom and a low-gain antenna mounted at the end of another boom. Two solar panels, 2.69m

Left: US officials with a model of the Pioneer Moon probe in 1958. Pioneer was to have entered orbit around the Moon, an extraordinary ambition for its time. However, it failed to reach the Moon.

(8.83ft) long with 19,800 silicon cells, provided 820W of power.

Mariner 10 astonished scientists who thought they were looking at the Moon rather than Mercury. Craters, ridges, lava-flooded areas, and the mountainous ring of an impact basin came into view. Other instruments detected temperatures ranging from -183°C to +187°C, and a metallic core, representing 80 per cent of the planet, 4885km (3035 miles) in diameter.

PIERCING THE CLOUDS OF VENUS

More than 12 attempts to explore Venus have failed, four of them before the first success by Mariner 2 in December 1962, which became the first spacecraft to explore any planet. The 203kg (447.5lb) spacecraft was launched in August 1962, and reached its target in four months. The craft was based around a bus with twin solar panels, 1.52m (4.9ft) long, each with 4900 silicon cells providing up to 222W of electricity. The craft was 3.02m (9.9ft) high, and supported a pyramid truss-frame on which six of the craft's seven experiments were mounted.

As the planet was shrouded in thick cloud, no camera was fixed to the craft. Radiometers aboard the craft produced the first shock when they indicated the surface temperature as 425°C. The carbon dioxide cloud cover – which Mariner found to be at its thickest between 80km (49.7 miles) and 56km (34.8 miles) altitude – was causing an extreme greenhouse effect. The veil of the

Far left: A montage of all the planets in the solar system which have been explored by space craft. Moving away from the Sun (smallest in picture) they are: Mercury, Venus, Earth, Mars, Jupiter, Saturn, Uranus and Neptune.

Below: Mariner 10 was the first, and so far, only spacecraft to explore the planet Mercury. The craft made three fly-bys of the planet closest to the Sun, and also visited Venus en route.

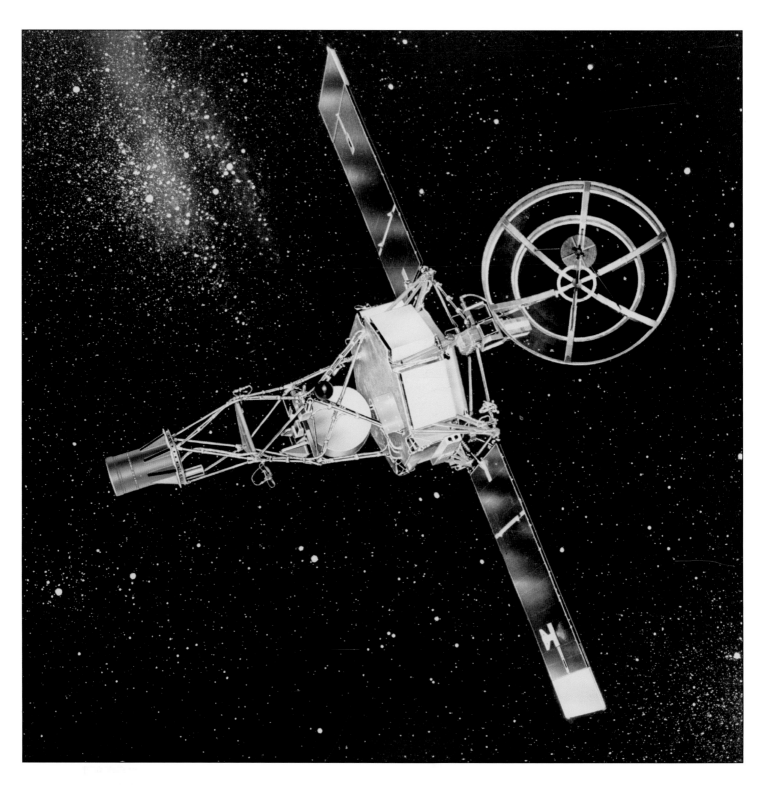

Above: The intrepid Mariner 2, the first space explorer launched in 1962, indicated that the surface temperature of Venus could be as high as 425ºC.

mysterious Venus had been lifted, revealing a hellish world, rather than the lush paradise which some had imagined.

The first spacecraft to penetrate the Venusian atmosphere was Venera 4, launched by the Soviet Union in June 1967. The craft probably did not survive to hit the surface under its parachute, as its transmissions ceased at an altitude of 27km (16.78 miles), when the atmospheric pressure had already reached about 22 times that of Earth's, and the temperature about 280˚C. The main discovery by Venera 4 was that the atmosphere consisted of 95

PIONEER VENUS 1

The Pioneer-Venus orbiter was the first spacecraft to carry a radar imaging system to penetrate the thick cloud cover of Venus and survey the surface of the planet.

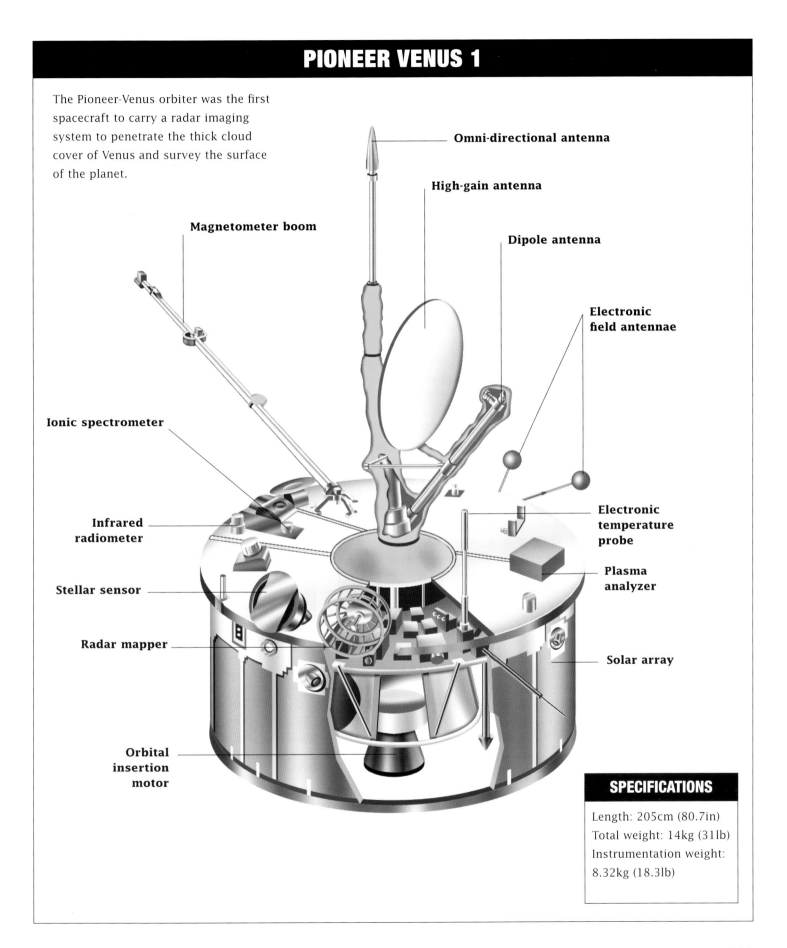

Omni-directional antenna

High-gain antenna

Dipole antenna

Magnetometer boom

Electronic field antennae

Ionic spectrometer

Infrared radiometer

Electronic temperature probe

Stellar sensor

Plasma analyzer

Radar mapper

Solar array

Orbital insertion motor

SPECIFICATIONS

Length: 205cm (80.7in)
Total weight: 14kg (31lb)
Instrumentation weight: 8.32kg (18.3lb)

Above: Landing probes from Venera 13 and 14 measured the chemical properties of the soil on Venus, showing it to be similar to basaltic granite found on Earth.

(3.3ft) in diameter, was attached by shock absorbers to a collapsible doughnut-shaped impact ring inflated with air, while above it was situated a 2.1m (6.88ft) diameter aerobraking disc. Parachutes were dispensed with, since experience had shown that the planet's thick atmosphere slowed the craft down considerably. Venera 9 came down on a 15-20° slope at the base of a hill, with conditions measuring 90 atmospheres and 460°C. The first images showed rocky surface and lighting conditions similar to a cloudy winter's day. Veneras 13 and 14 later returned the first colour images from the surface.

The first radar mapping of the planet was made by Pioneer Venus 1, which was launched in May 1978 and entered orbit around the planet in December. The design of the craft, 553kg (1219lb) in weight, 2.53m (8.3ft) in diameter and 1.22m (4ft) high, was a cylindrical bus which was spin-stabilized at 5rpm during its orbits of Venus. The 14,580 solar cells were mounted around its circumference, providing 312W of power. The craft's solid-propellant retrorocket placed it in a 24-hour orbit around Venus, with an eventual perigee of 150km (93.2 miles) and an apogee of 66,889km (41,563 miles).

The radar mapper allowed a topographical map of most of the surface to be made, between 73°N and 63°S, with a 75km (46.6 mile) resolution. This revealed an extraordinary scene of a surprisingly smooth surface, featuring two large continents, called Ishtar Terra and Aphrodite Terra, the size of Australia and Africa respectively, and a possibly extinct volcano called Maxwell Montes, 10.8km (6.7 miles) high. Meanwhile, Pioneer Venus 2 had also arrived at Venus, carrying four atmosphere probes which made a startling discovery – the Venusian atmosphere is almost totally composed of sulphuric acid.

The Soviet Union's exploration of Venus ended in June 1985 when two Vega

per cent carbon dioxide. Veneras 5 and 6 followed in 1969, and their capsules ruptured under the intense pressure after about 50 minutes of transmission.

Venera 7's capsule became the first to land and continue to tell the tale, and is rightly credited with the first landing on Venus. Venera 7 was launched in August 1970, and reached its destination in December of that year. The much-strengthened capsule, weighing 500kg (1102lb), indicated a temperature of 475°C and an atmospheric pressure equivalent to 90 Earth atmospheres. The first surface pictures of Venus were later taken by the descent vehicles of Veneras 9 and 10 in 1975.

The descent craft were more sophisticated and stronger than earlier landers. The central pressure vessel, 1m

VENERA 9 AND 10

Venera 9 and 10 orbited Venus in October 1975 and
also landed two craft on the surface which returned
the first TV pictures from the fiercely hot planet.

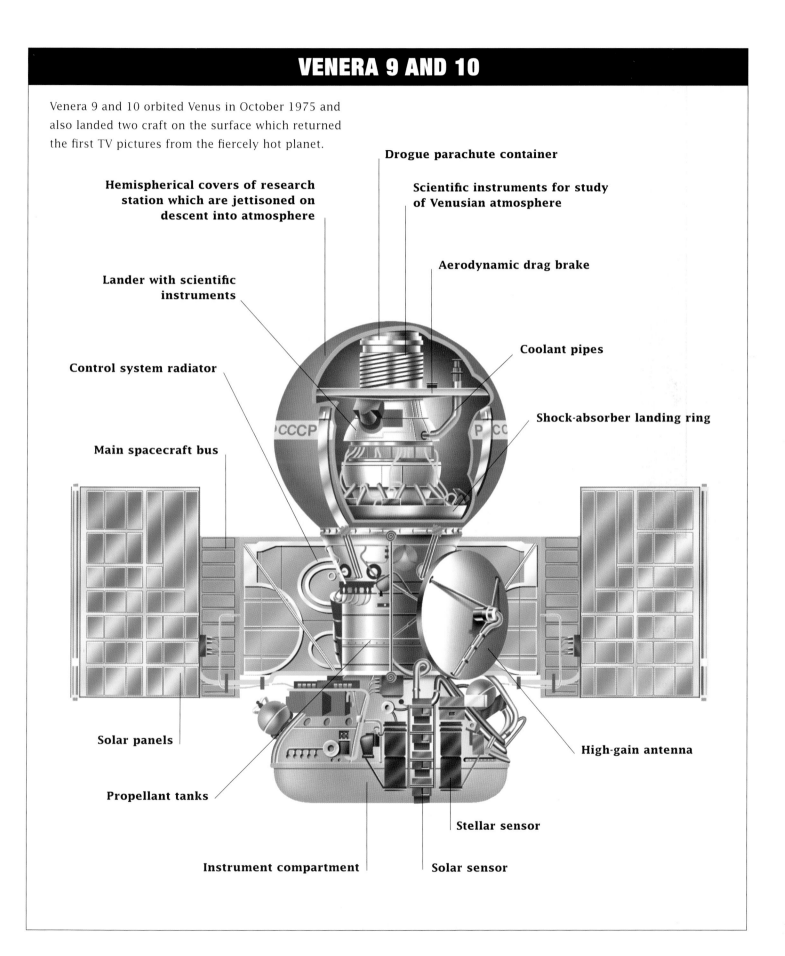

Drogue parachute container

**Hemispherical covers of research
station which are jettisoned on
descent into atmosphere**

**Scientific instruments for study
of Venusian atmosphere**

Aerodynamic drag brake

**Lander with scientific
instruments**

Coolant pipes

Control system radiator

Shock-absorber landing ring

Main spacecraft bus

Solar panels

High-gain antenna

Propellant tanks

Stellar sensor

Instrument compartment

Solar sensor

Below: Launched on 28 July, Ranger 7 carried a battery of six TV cameras on a lunar impact trajectory. Unlike Ranger 6, which failed at the moment the cameras were turned on, Ranger 7 was a triumphant success. The pictures revealed the surface of the Moon in startling detail, showing the topography of craters as small as 30m (100ft) in diameter.

spacecraft, en route to a rendezvous with Halley's Comet, dropped off two descent modules, weighing 1.5 tonnes and 2.39m (7.84ft) in diameter. Each carried standard Venera-like landers which, at 55km (34.2 miles) altitude, deployed a Teflon-coated plastic balloon, 3.54m (11.6ft) in diameter, inflated with helium. At the end of a 13m (42.7ft) tether was attached a three-section gondola which incorporated nine instruments which returned data.

The last major Venus mission did not begin until 1989, when the Space Shuttle Atlantis STS 30 deployed the Magellan Venus radar mapper in Earth orbit. This was duly dispatched to Venus by igniting an inertial upper stage (IUS) attached to it. The 3.44 tonne spacecraft, 6.46m (21.2ft)

high and 4.61m (15.12ft) wide, entered a polar orbit around Venus to begin an intensive and detailed global mapping mission which lasted three years. It provided images with a surface resolution of 120m (393.7ft), at a rate of 1852 per orbit, during which it took swathes of spectacular images between 17km (10.56 miles) and 28km (17.4 miles) wide, featuring Maxwell Montes, craters, mysterious pancake-like domes, and other features.

MISSIONS TO THE MOON

The first man-made object ever to hit the Moon was the Soviet Union's Luna 2, an aluminium-magnesium sphere, 1.2m (3.93ft) in diameter, and equipped with three simple instruments, some of which were mounted on protruding booms. The craft, weighing 390kg (860lb), remained attached to the upper stage throughout the rapid one-day flight, and plummeted to the Moon on 13 September 1959 at a speed of around 3.3km (2 miles) per second. Its transmissions ended abruptly on the edge of the Mare Imbrium, close to the crater Archimedes.

The next triumph was achieved by Luna 3, launched a month later. This 278kg (613lb) craft was 1.3m (4.27ft) long with a diameter of 1.19m (3.9ft), and carried solar cells on its body, a first for the Soviets. Its main instrument was a photographic-TV imaging system, which was called into action 65,000km (40,389 miles) above the Moon on 7 October, taking 29 images in 40 minutes, chiefly of the Moon's far side, hitherto unseen. The 35mm film was developed, fixed and dried by an on board processor and converted by a light beam into 1000 lines-per-image TV pictures which were transmitted to Earth, revealing to mankind something that they had never seen before – 70 per cent of the hidden face of the Moon. It was an enormous achievement, and marked one of the major milestones in the history of the Space Age.

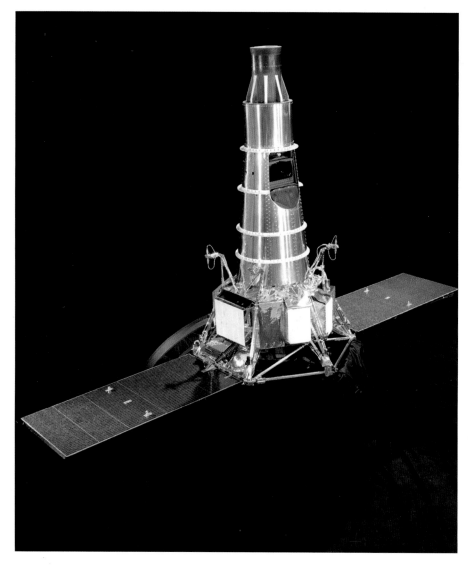

Left: Luna 3, which scored another triumph for the Soviet Union by making the first reconnaissance of the far side of the Moon in 1959.

The next major milestone was to be close-up images of the Moon. The 366kg (807lb) Ranger 7 spacecraft plummeted into the Moon's Sea of Clouds at 9316km/h (5789mph) on 31 July 1964, after transmitting 4316 pictures. The final image showing a mottled surface with hundreds of small craters, a thousand times better than anything seen before through a telescope on Earth. Ranger 7 was followed by the equally successful Rangers 8 and 9 in the series. The Rangers were based on a spacecraft bus similar to that used in Mariner 2. Mounted on it was the 1.5m

Above: The first spacecraft to impact the Moon, in September 1959, was the Soviet Union's Luna 2 spacecraft which remained attached to its upper stage.

(4.9ft) tapered, tower-like payload body, at the end of which was housed the 173kg (381lb) TV system, comprising six cameras. The images recorded on vidicons were scanned for TV transmission as they were taken.

The first soft landing on the Moon was achieved by the Soviet Union's Luna 9 in February 1966, but it was not strictly speaking "soft". The 100kg (220.5lb) surface capsule of Luna 9 travelled to the Moon attached to the main spacecraft bus and was equipped with a 1.5kg (3.3lb) TV camera. As the spacecraft plummeted to the surface, a 45.5 kilonewton rocket was fired to slow the descent, and at five metres above the surface, the 58cm (22.8in) diameter capsule was ejected and, like the

main craft, impacted on the Moon at 22km/h (13.7mph). The capsule came to rest and started transmitting, opening four "petals" to expose the TV camera, which worked rather like a fax machine for transmission. The camera rotated 360° to produce a 6000-line panorama in 1 hour 40 minutes, stretching only 1.5km (0.93 miles) into the distance. The powdery soil was covered with small stones of various sizes, proving that the lunar dust – at least in the Ocean of Storms – was not deep.

The first true soft landing, however, was made by the US Surveyor 1 which touched down with the aid of a retrorocket in June 1966. Altogether, five Surveyors made soft landings on various parts of the Moon, returning a wealth of imagery to assist

LUNAKHOD 2

Luna 21 soft-landed inside Le Monnier crater near the eastern rim of the Sea of Serenity on 16 January 1953. The first period of lunar exploration began on 17–18 January when Lunakhod 2 moved off from the landing site in a south-easterly direction over basalt lava, negotiating craters and boulders. Panoramic pictures received on Earth clearly showed the surrounding scene, including mountains bordering the Sea of Serenity.

SPECIFICATIONS

Weight: 840kg (1852lb)
Dimensions: 221cm (87in)
Wheel diameter: 51cm (21in)

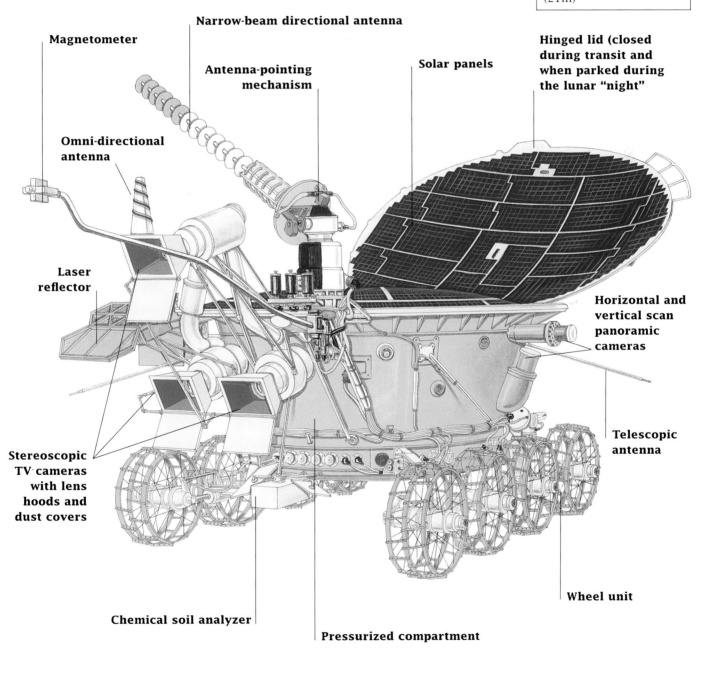

Magnetometer

Narrow-beam directional antenna

Antenna-pointing mechanism

Solar panels

Hinged lid (closed during transit and when parked during the lunar "night"

Omni-directional antenna

Laser reflector

Horizontal and vertical scan panoramic cameras

Stereoscopic TV cameras with lens hoods and dust covers

Telescopic antenna

Wheel unit

Chemical soil analyzer

Pressurized compartment

Above: A model of Luna 9 which made the first "soft" landing on the Moon in January 1966.

Orbiter, rather than a genuine project in its own right.

The first return to Earth of moondust was achieved by the Soviet Union, using an unmanned spacecraft, Luna 16, which was also the Soviet's first true soft lander. Luna 16 landed on the Moon in September 1970 and incorporated a lunar collector which scooped up soil and deposited it inside the capsule positioned on top of the ascent stage. The capsule, 0.5m (1.64ft) in diameter and weighing 39kg (86lb), eventually parachuted to Earth carrying 101g (3.56oz) of lunar material, including an intriguing glass sphere from the Sea of Fertility. By this time, of course, Apollo missions had returned with pounds of rock.

Before the Apollo programme, the Soviet Union deployed an unmanned Lunakhod rover which flew aboard Luna 17 in November 1970. This Heath Robinson-like machine, looking like a bathtub on eight wheels, weighed 756kg (1667lb) and measured about 1.35m (4.43ft) high and 2.15m (7ft) in length. It was equipped with science instruments and cameras, and was operated remotely from Earth. Each titanium and fine mesh wheel, 51cm (20in) in diameter, was operated by its own electric motor on the hub. The amazing spacecraft operated for 11 days, travelling a total of 10km (6.21 miles) across the Sea of Rains. Lunakhod 2 was launched in 1973.

THE QUEST FOR MARS

While Russia achieved significant milestones in the exploration of Venus, its attempts to explore the planet Mars were disappointing, with only the Mars 5 orbiter being a total success. In contrast, the US had been achieving spectacular success until late 1996 when it lost both the Mars Climate Orbiter and Mars Polar Lander. This prompted a major re-think about the future of Mars exploration: their loss resulted in the cancellation of the next lander.

Apollo managers to select sites for lunar landings, complemented by a fleet of five amazingly successful Lunar Orbiters. The Soviet Union, however, also had the distinction of being the first in lunar orbit, with Luna 10 in April 1966. However, it seemed to be a rushed effort to beat Lunar

MARINER 9

Mariner 9, the first spacecraft to enter orbit around Mars, swung into orbit on 13 November 1971 after successfully completing a braking manoeuvre. Instruments designed to explore the planet included narrow-angle and wide-angle TV cameras on a scan platform, an infrared interferometer spectrometer to measure gases, particles and temperatures on and above the surface; an ultra-violet spectrometer to identify gases in the upper atmosphere; and an infrared radiometer to measure surface temperatures. At the time of the spacecraft's arrival, a dust storm hid many features of the planet. When the dust cleared, its camera made startling discoveries – an equatorial canyon bigger than Earth's Grand Canyon, volcanoes, and features that resembled dried-out river beds.

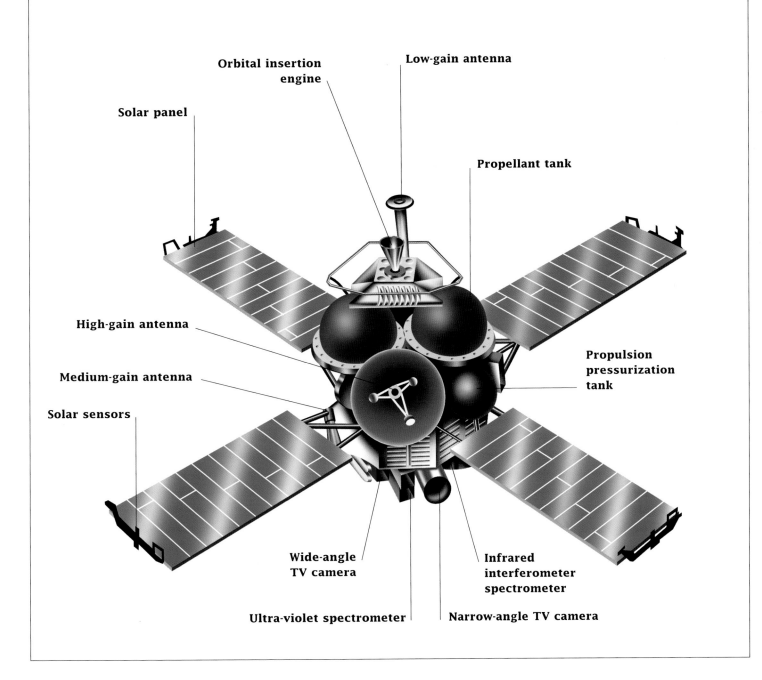

Orbital insertion engine

Low-gain antenna

Solar panel

Propellant tank

High-gain antenna

Propulsion pressurization tank

Medium-gain antenna

Solar sensors

Wide-angle TV camera

Infrared interferometer spectrometer

Ultra-violet spectrometer

Narrow-angle TV camera

VIKING ORBITER

Two Viking spacecraft made the first landings on the Martian surface, while two mother ships surveyed the planet in great detail from orbit.

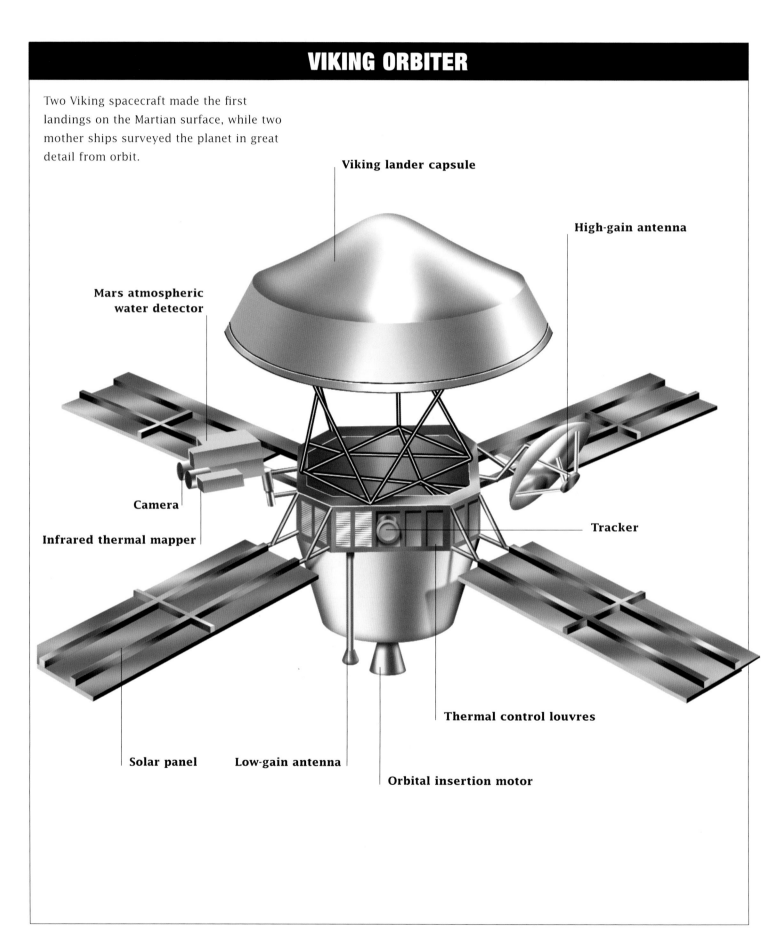

Viking lander capsule

High-gain antenna

Mars atmospheric water detector

Camera

Infrared thermal mapper

Tracker

Thermal control louvres

Solar panel

Low-gain antenna

Orbital insertion motor

Mariner 4 was launched in November 1964, and reached the planet the following July, revealing that much of it appeared to be like the Moon, covered in craters. The spacecraft bus was similar to that which later flew on Mariner 10 to Mercury, measuring 1.38m (4.32ft) in diameter and 0.45m (1.48ft) in height. It was equipped with four solar panels, comprising 7000 solar cells which produced 700W of power, and a single TV scanning camera programmed to return just 21 images as it passed the Red Planet at its closest distance of 9600km (5965 miles).

The images covered a swath from 37°N to 55°S, and each comprised 40,000 elements, transmitted as numbers according to the amount of light. Other instruments returned data on the atmosphere. Many of the images were rather fuzzy and it was not until Mariner 11 that the surface would be seen in relatively fine detail, when several craters in a region called Atlantis were revealed. This morsel of Martian information whetted the appetite for the missions of later Mariner craft.

The 565kg (1246lb) Mariner 9 was launched in May 1971, and became the first Mars orbiter in November, entering an 80° orbit which would greatly increase the coverage to 70 per cent of the entire surface of the planet. It would also obtain further data from a suite of instruments on the Mariner 4-like spacecraft bus equipped with a retrorocket package, increasing its height to 2.3m (7.54ft). Initial images indicated that Mariner had arrived during a huge dust storm which obliterated the surface until January 1972. The craft

managed to take a number of images of the Martian moons, Phobos and Deimos, showing them to be pockmarked by craters and probably captured asteroids.

As the dust cleared, Mariner's panorama proved to be spectacular, revealing a diverse surface, indicated by earlier Mariner 6 and 7 fly-bys, of more than just craters. There also appeared to be dry river beds, huge valleys, including the 4000km (2485 mile) long, 100km (62 miles) wide, Valles Marineris, and volcanoes, including the 25km (15.5 miles) high Olympus Mons. Mariner 9's pictures were also used to select the landing sites for the Viking programme.

Two Viking craft were launched in August and September 1975, and reached the surface of Mars on 20 July and 3 September 1976 respectively. The two identical Vikings were spectacular successes. They also included two orbiters which continued the intensive survey of Mars begun by Mariner 9. The landers were encased in a lens-shaped 366kg (806lb) aeroshell which plunged into the upper Martian atmosphere, and was protected against temperatures of 1500°C. A parachute with a diameter of 16.2m (53ft) opened at about 5km (3.1 miles) altitude. The aeroshell was jettisoned and the lander's legs deployed, and at about 1.4km (0.87 miles) above the surface, the craft's three throttleable engines and four thrusters ignited, slowing the craft for a touchdown at about 2.4m (7.9ft) per second.

The craft were basically six-sided aluminium-titanium buses, which housed

Below: This spectacular panorama of Chryse Panitia was returned by Mars Pathfinder soon after its Independence Day landing.

FIRST PLANETARY EXPLORATIONS

Date	Vehicle	Country	Event
MOON			
12 September 1959	Luna 2	USSR	Launch
13 September 1959	Luna 2	USSR	Impact on moon at 30°N 0°Lat. First object to reach moon's surface.
4 October 1959	Luna 3	USSR	Launch
6 October 1959	Luna 3	USSR	Flight to within 6000km (3,750 miles) of Moon's surface. First images of 70 per cent of far side of Moon.
28 July 1964	Ranger 7	US	Launch
31 July 1964	Ranger 7	US	Impact on moon at about 10.7°S, 20.7W. First high-resolution images of surface before impact.
31 January 1966	Luna 9	USSR	Launch
3 February 1966	Luna 9	USSR	Landing at 7.13°N, 64.37°W. First"soft" landing and images from surface.
31 March 1966	Luna 10	USSR	Launch
3 April 1966	Luna 10	USSR	Entered lunar orbit. First lunar orbiter.
10 November 1970	Luna 17	USSR	Launch
17 November 1970	Luna 17	USSR	Landing at 38.28°N, 35°W. Lunokhod is first lunar rover.
VENUS			
27 August 1962	Mariner 2	US	Launch
14 December 1962	Mariner 2	US	Fly-by of 34,827km (21,766 miles). First Venusian explorer.
17 August 1970	Venera 7	USSR	Launch
15 December 1970	Venera 7	USSR	First transmissions from the surface at 5°S 351°Lat
8 June 1975	Venera 9	USSR	Launch
22 October 1975	Venera 9	USSR	First transmission of images at 32°N 291°Lat. First Venus orbiters.
MARS			
28 November 1964	Mariner 4	US	Launch
15 July 1965	Mariner 4	US	Flight to within 9600km (6000 miles) of Mars. First fly-by and images.
30 May 1971	Mariner 9	US	Launch
14 November 1971	Mariner 9	US	Entry into orbit. First Mars orbiter
20 August 1975	Viking 1	US	Launch
19 June 1976	Viking 1	US	Entry into orbit
20 July 1976	Viking 1	US	Lander touchdown at 22.483°N, 47.94°W. First Mars landing, surface pictures and sample analysis.
2 December 1996	Mars Pathfinder	US	Launch
4 July 1997	Mars Pathfinder	US	Landing on Mars. Deployment of first Mars roving vehicle, Sojourner.
JUPITER			
3 March 1972	Pioneer 10	US	Launch
5 December 1973	Pioneer 10	US	Flight to within 130,000km (81,250 miles) of Jupiter. First exploration of Jupiter and close up images.
13 October 1989	Galileo	US	Launch

Date	Vehicle	Country	Event
7 December 1995	Galileo	US	Entry into orbit of Jupiter. First Jupiter orbiter and first capsule penetration of atmosphere.

SATURN

Date	Vehicle	Country	Event
6 April 1973	Pioneer 11	US	Launch
1 September 1979	Pioneer 11	US	Flight to within 20,900km (13,062 miles) of Saturn. First Saturn fly-by

MERCURY

Date	Vehicle	Country	Event
3 November 1973	Mariner 10	US	Launch
29 March 1974	Mariner 10	US	Flight to within 703km (439 miles) of Mercury
21 September 1974	Mariner 10	US	Fly-by of 48,069km (30,034 miles)
16 March 1975	Mariner 10	US	Closest proximity to Mercury of 327km (232miles) First and so far only exploration of Mercury. In also passing Venus, Mariner 10 is first spacecraft to explore two planets.

URANUS

Date	Vehicle	Country	Event
20 August 1977	Voyager 2	US	Launch
24 January 1986	Voyager 2	US	Flight to within 71,000km (44,375 miles) of Uranus. First and so far only exploration of Uranus. First spacecraft to explore three planets (including passing of Jupiter and Saturn).

NEPTUNE

Date	Vehicle	Country	Event
25 August 1989	Voyager 2	US	Flight to within 5016km (3135 miles) of Neptune. First and so far only exploration of Neptune. First spacecraft to explore four planets.

COMETS

Date	Vehicle	Country	Event
12 August 1978	ICE/ISEE 3	US	Launch
11 September 1985	ICE/ISEE 3	US	Flight to within 7862km (4913 miles) of comet Giacobini Zinner on First comet explorer
2 July 1985	Giotto	Europe	Launch
14 March 1986	Giotto	Europe	Flight through the coma of Halley's Comet, to within 606km (378 miles) of nucleus. First exploration of cometary coma.

ASTEROIDS

Date	Vehicle	Country	Event
18 October 1989	Galileo	US	Launch
29 October 1991	Galileo	US	Flight to within 1604km (1002 miles) of Gaspra. First asteroid exploration and close-up images.
17 February 1996	NEAR	US	Launch
14 February 2000	NEAR	US	Entry into orbit around Eros. First asteroid orbiter.

Right: Valles Marineris, a huge
canyon that spreads across
the Martian surface was
photographed in great detail
by the Viking orbiters.

external instruments, including a meteorology boom, with a maximum height of 2.1m (6.88ft). Each of the three landing legs had 0.3m (0.98ft) diameter footpads, and the craft was powered by two radioactive, plutonium oxide, radioisotope, thermoelectric generators. Each carried a 3m (9.84ft) robot arm with a scoop with which to collect soil. The soil was deposited into an internal laboratory comprising a biology distributor, gas chromatograph mass spectrometer, and an X-ray spectrometer. Ingenious methods to detect signs of life on Mars were employed, including heating the samples, and adding water and nutrients.

However, no conclusive proof of the existence of life was made by either of the landers which returned wonderful images of the rust-stained surface and the far horizon, a pink sky and carbon-dioxide frost on the surface. The atmosphere is almost entirely carbon dioxide, with an atmospheric pressure on the surface of 7.6mb, dropping by 30 per cent in the winter. The temperature was about -33˚C in mid-afternoon, and wind speeds of up to 51km/h (40mph) were detected. The landings by Vikings 1 and 2 are considered to be two of the major milestones in space exploration, and were not followed until 1997, when Mars Pathfinder captured the imagination of internet users worldwide, especially with its little rover, Sojourner. The mission cost $265 million, compared with the $1 billion spent on the Vikings, a reflection of how limited the space budgets of today have become.

Launched in December 1996, the mission was one in a Discovery series to

Above: The 9kg (19.8lb) Sojourner rover, which was deployed from the Mars Pathfinder spacecraft in July 1997, was controlled by an Earth-based operator.

Above: Voyager 2 explored four planets — Jupiter, Saturn, Uranus and Neptune — during its extraordinary odyssey through interplanetary space and beyond.

demonstrate low-cost exploration of the Martian surface. Sojourner was a six-wheeled vehicle, the size of a microwave oven, mounted on a "rocker-bogie" suspension. The rover was stowed on the Mars Pathfinder lander. After the bouncing landing of the main craft on 4 July 1997, encased in inflated helium balloons, the 9kg (19.84lb) rover rolled down a deployment ramp, controlled by an Earth-based operator, who later used images obtained by both the rover and lander for navigation. The landing site was a large outwash plain near Chryse Planitia in the Ares Vallis region of Mars, at 19.33˚N 33.55˚W. This is one of the largest outflow channels on Mars, and is the possible result of a huge flood of water over a short period of time, which flowed into the northern Martian hemisphere.

TRIPS TO THE GIANT PLANET

The first spacecraft to make the trip to Jupiter – the largest planet in the solar system – was Pioneer 10, launched in March 1972 and which reached its destination in December 1973. The 258kg (569lb) Pioneer was based on a spacecraft bus of hexagonal shape to contain the electronics, and was topped by a large antenna, 2.74m (8.9ft) wide, made of an aluminium honeycomb. The craft did not have solar panels since the Sun's power would be so low by the time it reached Jupiter – 4 per cent of that received by the Earth. Instead, mounted at the end of two of the booms on the craft were two SNAP radioisotope thermo-electric generators, using plutonium 238 as fuel, and providing 140W. A third boom, 6.55m (21.5ft) long, carried a magnetometer, while the other 10 instruments were mounted on the bus itself, and included an array of telescopes and detectors, as well as a TV camera.

Pioneer also carried a plaque on its side which depicted two human beings and their position in the solar system, in case the craft is one day found by another civilisation, if one exists. More than 300 images of the planet were received, and of much better resolution than those seen on the best telescopes on Earth. They included the Great Red Spot, and the damage caused by the large radiation belt that was discovered surrounding the planet. Pioneer 10 was followed to Jupiter by its sister ship, Pioneer 11, launched in April 1973 and reaching the planet in December 1974. This spacecraft was the first to use a planet as a "slingshot", flying by the south of Jupiter and being whipped northward, eventually in the direction of Saturn by the gravity-assist manoeuvre.

The next visits to Jupiter were made by Voyagers 1 and 2 in 1979, and the picture

Left: The volcanically active, sulphurous moon of Jupiter, Io, spews material into space, as seen on the horizon. The surface of Io is being torn by the forces of the giant planet it orbits.

CASSINI

The Cassini project represents an important collaboration between NASA (who provided the Saturn orbiter) and the ESA (who constructed the atmosphere probe). Saturn is known to have 22 moons, ranging in size from a few kilometres to over 5000km (3000 miles). Titan is the largest moon which, with a diameter of 5140km (3194 miles) is larger than the planet Mercury. The Huygens probe is due to detach from the Cassini spacecraft and descend through Titan's cloud belt in 2004.

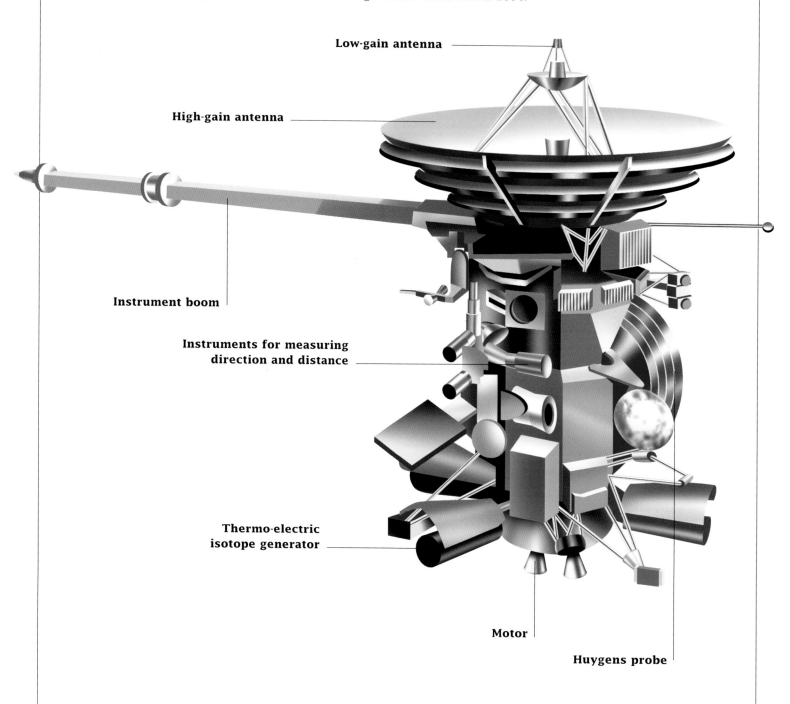

Low-gain antenna

High-gain antenna

Instrument boom

Instruments for measuring
direction and distance

Thermo-electric
isotope generator

Motor

Huygens probe

alternative systems to their fullest.

The main payload of Galileo was the 339kg (747lb) instrumented descent probe which was released some 80 million kilometres (50 million miles) from Jupiter in July 1995. It plunged into the cloud deck in December, confirming an intense radiation belt around the planet at 50,000km (31,000 miles) distance. Closer to Jupiter, it detected only one cloud layer, 640m (2000ft) per second winds, and traces of organic compounds in the clouds after its high-g, high-temperature entry into the atmosphere and subsequent parachute deployment. The probe survived for one hour 15 minutes until it was destroyed by the intense pressures. Galileo itself entered orbit on 8 December 1995, beginning an epoch-making exploration of the planet and its highly individual moons, particularly the volcanic Io and ice-covered Europa. The exploration continued into 2000.

VOYAGES TO THE RINGED PLANET

Saturn is also known as the ringed planet, after its spectacular system which has fascinated astronomers for more than 400 years since the first telescope was pointed in its direction. The first close-ups of Saturn's rings were taken by Pioneer 11, in September 1979, which was later followed by Voyagers 1 and 2. Saturn's ring system was found to consist of thousands of individual rings, made up of ice and small rock particles of up to 1m (3.28ft) in diameter, all held together in orbit by the gravitational pull of the planet. The spacecraft also took startling images of the moons of Saturn, including the enigmatic Titan. This moon was found to be covered with a nitrogen-thick atmosphere which planetary scientists and biologists found tantalising, leading to the development of a future Saturn orbiter and Titan lander.

During its journey past Saturn, Voyager 2 was able to use the planet's gravity as a "slingshot", sending it en route to a

quality from Jupiter had greatly improved since Pioneer. The most spectacular images were of Jupiter's four largest moons, Io, Europa, Ganymede and Callisto.

The next spacecraft to Jupiter was called Galileo. It was the first to orbit the planet and the first to deploy a descent capsule to plunge into its thick cloud-deck. The spacecraft was deployed from the Space Shuttle Atlantis in October 1989 and, en route to Jupiter, flew past Earth twice in December 1990 and 1992, and past Venus in February 1990. The resulting gravity-assisted flight path and velocity changes also enabled Galileo to become the first spacecraft to fly-pass asteroids, Gaspra in October 1991 and Ida in August 1993, taking the first close-up pictures of these main-belt bodies. Galileo was a large spacecraft, weighing 2.22 tonnes, and incorporated a high-gain antenna which, however, failed to deploy properly. This reduced the amount of data that could be transmitted back to Earth, although engineers found ways to use the

GIOTTO

The Giotto probe was launched by Ariane on 2 July 1985 for a rendezvous some eight months later with Halley's Comet. The spin-stabilized spacecraft carried 10 scientific instruments to study the comet: a camera; neutral-mass, ion-mass, and dust-mass spectrometers; a dust-impact detector system; two plasma analyzers; a magnetometer; an energetic particles experiment; and an optical probe.

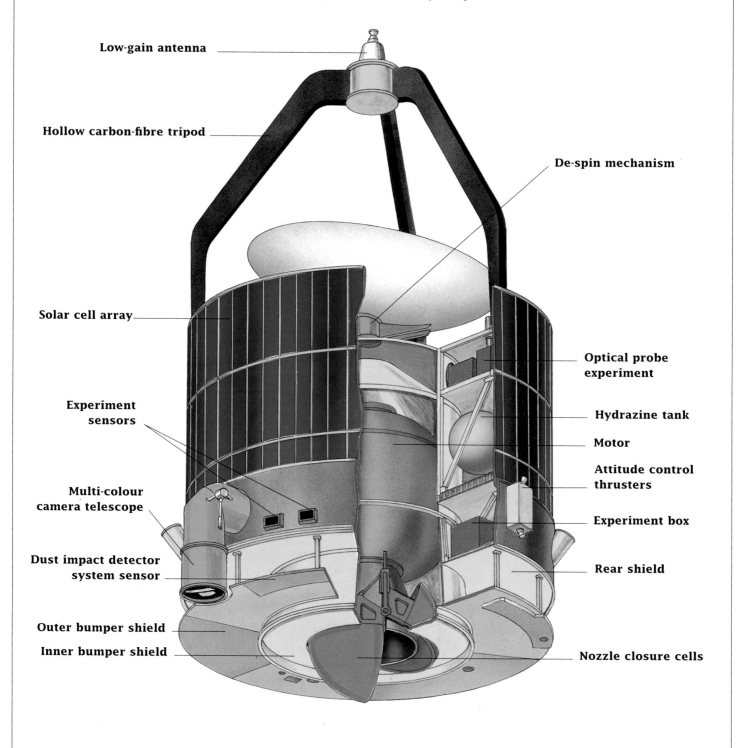

Low-gain antenna

Hollow carbon-fibre tripod

De-spin mechanism

Solar cell array

Optical probe experiment

Experiment sensors

Hydrazine tank

Motor

Attitude control thrusters

Multi-colour camera telescope

Experiment box

Dust impact detector system sensor

Rear shield

Outer bumper shield

Inner bumper shield

Nozzle closure cells

rendezvous with Uranus and Neptune. The 825kg (1819lb) Voyager 2's appearance was dominated by an antenna, 3.66m (12ft) in diameter, for both X- and S-band communications. Beneath the antenna sat the spacecraft bus, from which three booms and a further antenna extended. One of the booms held most of the instruments, including wide and narrow-angle TV cameras. Both Voyager 2 and Voyager 1 carried a 30cm (11.8in) diameter gold-plated copper disc, similar to a gramophone record, with a needle and playing instructions mounted on the side of the bus, which contained recordings of natural Earth sounds, 90 minutes of music, 115 analogue pictures and greetings in 60 languages.

Uranus proved to be a disappointment visually, with little variation in its greenish hydrogen-helium-methane atmosphere. However, one of its moons, Miranda, certainly caught the eye, appearing to have been a moon which had broken apart and then fused together again. Neptune's encounter featured the planet's "scooter"

Right: The remarkable Giotto spacecraft which survived a "suicide mission" flying through the coma of Halley's Comet, and went on to explore another comet.

Above: Pioneer 10, the legendary interplanetary explorer, was launched in 1972 and made the first exploration of Jupiter before passing out of the solar system.

clouds racing around the upper atmosphere at 2000km/h (1243mph), and the moon Triton, at –235°C, the coldest known object in the solar system.

THE COMET FLEET

The year 1986 heralded the long-awaited return to the skies of Halley's Comet. For the first time in its regular visits (which occur every 76 years) to the vicinity of the sun, during its long and lonely orbit into deep space, when it heats up and sheds material like an aspirin fizzing in water, the spectacle was to be watched from close quarters by an international fleet of spacecraft from Russia, Japan, the US and Europe.

The star of the show was Europe's Giotto, a probe designed to fly right through the coma of the comet in one of the most remarkable and rather unheralded flights in history. Giotto was a spin-stabilized, drum-shaped craft built in the UK. It weighed 960kg (2116lb) at launch, was 1.867m (6.13ft) in diameter and 2.848m (9.34ft) high. It was spin-stabilised at 4rpm during the actual encounter with the comet, and was protected by a "sandwich" shield of aluminium and a specialized composite material called Kelvar, set 23cm (9in) apart from each other. The shield was required in order to protect the craft from impacts of comet dust and particles as the craft travelled at a speed of 68km (42 miles) per second through the comet – fast enough to make a trans-Atlantic trip in around 1.5 minutes. The craft was supplied with 190W of electricity from over 5000 solar cells around its body, and was equipped with 10 experiments which included camera and dust analyzers and detectors.

The remarkable Giotto shot through the upper part of the coma of Halley's Comet on 13 March and discovered that 10 tonnes of water molecules and three tonnes of dust were being thrown out of the comet every second, as Giotto's shield took a battering which could be heard in transmissions to Earth. The best image at a distance of 18,000km (11,185 miles) showed that the nucleus of the comet was 15km (9.3 miles) long and 7–10km (4.3 to 6.2 miles) wide, and had two large jets of dust and gas erupting from what appeared to be cracks in the undulating surface of "hills" and "valleys".

THE ASTEROID ORBITER

Although the Galileo spacecraft had passed close to main-belt asteroids Gaspra and Ida en route to its target, Jupiter, the first spacecraft to explore an asteroid at close quarters on a dedicated mission, and eventually to enter orbit around an asteroid in February 2000, was the Near Earth Asteroid Rendezvous craft (NEAR). The asteroid in question was Eros, not a main-belt asteroid between Mars and Jupiter, but one which, in its elliptical orbit around the Sun, comes close to Earth. NEAR was launched in February 1997, and in June flew close to the asteroid Mathilde, aiming to enter orbit around Eros in February 1999. However, computer and engine errors caused the mission to be aborted, but NEAR was rescued and made it safely to Eros a year later than planned.

Part of the NASA Discovery programme,

NEAR weighed 805kg (1775lb), set on a 1.5m (4.9ft) high hexagonal bus, with four solar panels extended from its base, using gallium arsenide cells to provide 1.8kW of power. NEAR entered a lower orbit around Eros, between 366km (227 miles) and 200km (124 miles). NASA planned to have moved NEAR to as close as 50km (39 miles) by May 2000, where it would remain for several months before being manoeuvred into a deeper orbit again as part of the comprehensive imaging and monitoring process.

NEAR's orbit of Eros is planned to be reduced still further, to enable it to fly to within a few kilometres of the surface, and possibly even touching it, at very slow speed, thereby creating a "divot" in its surface. In addition to a CCD visible imager, NEAR carries an X-ray and gamma-ray spectrometer to map the elemental composition of the asteroid, and a laser altimeter. Initial images showed evidence of a layered structure, which may indicate that the 33km (20.5 mile) diameter asteroid is a remnant of a larger parent body which broke apart. They also showed a higher density of craters than was observed during fly-bys of main-belt asteroids.

Below: A montage of images taken by the NEAR spacecraft, showing the rotation of the asteroid Eros.

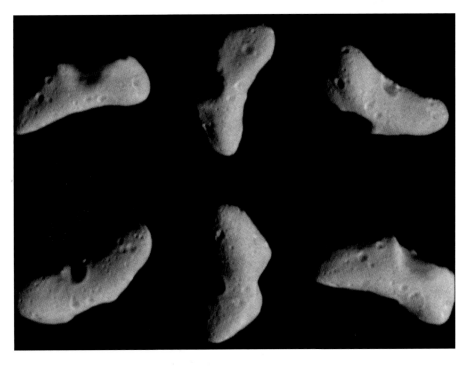

SPACE VEHICLES OF THE FUTURE

———

Predicting the future of space exploration and how mankind might exploit available resources in space is fraught with difficulties. Media expectations and corporate oversell have created an atmosphere of doubt and uncertainty around the value of space travel and exploration. To many, the Apollo programme, and the successful landing of men on the Moon was the culmination of man's efforts in exploring space.

The era of almost limitless budgets for space programmes was well and truly over by the end of the 1970s. Since then, space scientists and engineers have been hard pushed to justify the budgets required to support further development. Even the fully re-usable and much-heralded Space Shuttle follow-on vehicle, the VentureStar, may now never fly.

Left: An artist's impression of a futuristic mission involving the manned exploration of the surface of Mars, the "Red Planet".

The existing Space Shuttle will still be flying in 2015 with several improvements, such as new liquid boosters, while new vehicles like the VentureStar are still a long way from being realized. Indeed, several new launchers now under development, such as the Atlas V and Delta IV, are still based on the technology of the Cold War missile.

In the field of space exploration several near-term spacecraft, including Rosetta, which is due to be launched in 2003 to attempt the first landing on a comet, are fully funded. Other exploratory craft are already en route to their destinations. These craft include Cassini-Huygens which will reach Saturn in 2004, and Stardust, which is intended to bring samples of comet dust back to Earth.

As for the exploration of Mars, the failure of two spacecraft in October and December 1999 will inevitably prompt a radical re-think about NASA's programmes. The ultimate aim is to return samples of Mars to Earth, a target which had been set at 2005 , but is now likely to be much later. One unmanned landing mission has already been cancelled by NASA.

New space technologies, such as spacecraft artificial intelligence and methods of propelling spacecraft at greater speeds through the solar system, are being demonstrated as part of NASA's New Millennium programme by a spacecraft called Deep Space 1, which is already in deep space.

It is perhaps easier to predict trends in space applications, such as communications and navigation. Communications satellites will become increasingly important in the future, by serving users with instant access, from anywhere to anywhere. It will support personal communications needs with access to the internet and related services. The satellite-based Global Positioning System will also be a vital part of our daily lives, eventually helping us to drive to our destination, and avoid traffic congestion.

With regards to manned flight to Mars, it is still the realm of science fiction and is unlikely to happen before 2020 at the earliest, unless the unique combination of factors which gave birth to Apollo – technology, world tension and political will – arise once more. Space tourism may become possible, but it will still cost a million dollars or more for any would-be space traveller.

Finally, the future of space, especially human space travel, whether it is a Shuttle flight to the International Space Station, or a tourist hop into space, will depend a great deal upon safety. Flying into space is risk, and it is an extraordinary achievement

Below: Mobile and multi-media communications satellites will represent the largest business growth area of the future.

Left: Modules attached to the ISS. Each major component will require a Space Shuttle flight, of which as many as 40 may be needed to complete construction.

that in over 200 manned space flights, there have been only three fatal flights, including the tragic launch failure of the Shuttle in 1986. It must be accepted that another accident will happen.

The loss of a second Space Shuttle or any other manned space vehicle, however tragic an event, would be a benchmark, of sorts, against which future progress would be evaluated. Either an accident will be accepted as worth the risk, and the Shuttle and other vehicles will continue to fly, or there will be an outpouring of emotion, a Presidential Commission appointed to investigate it, and the programme grounded for years for redesigns.

IMPROVING THE SPACE SHUTTLE

The durable Space Shuttle fleet may still be flying in 2015, before it is succeeded by a new space vehicle which has yet to be designed. Although the Space Shuttle may look like the same vehicle which first flew in 1981, many critical improvements have already been made to it, and will continue to be added. The first major innovation is the so-called "glass cockpit". During the dynamic launch of the Space Shuttle, even

Right: The Space Shuttle orbiter docks with the ISS and the Canadian robot arm unloads the cargo from the payload bay. The arm is part of the ISS robotic manipulator system.

Far right: The new "glass cockpit" of the Space Shuttle made its debut on the orbiter Atlantis in 2000. Eventually, each Shuttle orbiter will be fitted with the new system.

milliseconds can make a significant difference. These glass cockpits will enable a crew to react faster and be in a better position to avoid disaster. NASA will eventually fit the new Boeing 777-style glass cockpit to all its orbiters, and has already been installed in the Space Shuttle Atlantis. Scores of outdated electro-mechanical cockpit displays such as cathode-ray tube screens, gauges, and instruments will give way to 11 full-colour flat panel screens. These screens provide Shuttle crews with easy access to vital information via the two and three-dimensional colour graphic and video capabilities of its on-board information management system.

Not only does the new system improve crew/orbiter interaction with the easy-to-read graphic portrayals of key flight indicators, such as craft-position display and Mach speed, but it also reduces the high cost of maintaining obsolete systems. Each display unit measures about 20cm (7.9in) square, weighs 8kg (17.6lb) and uses 67W of power, with a screen resolution of 172dpi. The hardware consists of the eleven identical full-colour, liquid-crystal

multifunction display units (MDU). Four of these directly replace the four monochrome units previously used. Two MDUs each replace the commander (CDR) and pilot (PLT) flight instruments; one MDU replaces the on-orbit manoeuvring instruments at the aft flight deck; and the remaining two MDUs replace the CDR and PLT status displays. The command and data entry keyboards, as well as the rotational and translational hand controllers and most of the other cockpit switches, remain unchanged.

Other planned Shuttle improvements include electric auxiliary power units which will replace the existing hydrazine-powered units which are extremely expensive to maintain. The original oxygen-hydrogen fuel cells used to generate electricity may be replaced by more powerful proton-exchange-membrane fuel cells, while a Space Shuttle Main Engine Advanced Health Management

system may be introduced to increase safety and reduce turnaround costs. Another planned upgrade is to switch the Shuttle's main propulsion-system propellant valve from pneumatic to electromechanical actuation. Other upgrades include more durable thermal-protection-system tiles for the underside of the orbiter, and changes to the main landing-gear tyres and improved abort systems. Each individual improvement may appear to be of little importance, but as a whole, the advanced Shuttle systems will enable it to fly for many more years to come.

One improvement that will make the Shuttle appear very different is the proposed liquid fly-back booster. The twin solid rocket boosters (SRB) of the Shuttle are effective, but offer little in the way of a safety margin, nor are they the most versatile flight-control options. Though not yet fully funded, the SRBs may be

Right: The original VentureStar design in 1996 featured a streamlined appearance, and an internal cargo bay. The next flight of the X-33 is scheduled for 2001.

VENTURESTAR

The experience of the VentureStar programme, and the technologies that it proposes to use will provide a good indication as to the type of Shuttle follow-up that will be needed, and the difficulties to be encountered. The VentureStar was introduced by NASA and its proposed builder Lockheed Martin in 1996 with the impression that passengers would be flying in space within ten years. There was an expectation that the VentureStar would start flying commercial missions in 2004, at a rate of 20 a year, and at a cost almost a tenth of that of the Space Shuttle. The reality was that only a scale-model technology-demonstrator vehicle, the X-33, was to be built, first flying in March 1999. A fully-fledged VentureStar would cost billions of dollars to develop, and many technological hurdles would have to be overcome before it became a reality.

The inevitable delays to the X-33 programme alone, which have resulted in a projected first flight in 2001, only serve to demonstrate that the step from the Shuttle to an airliner-type space vehicle is

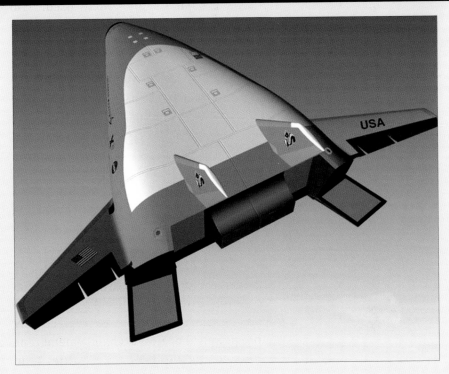

so great that an interim or complementary Shuttle follow-up vehicle may be developed at first. While the X-33 is regarded more as an advanced technology demonstrator, rather than a VentureStar prototype, the VentureStar itself has been extensively redesigned.

Above: The key technology of the X-33 is the rear-mounted linear aerospike engine, which boosts the craft into sub-orbital trajectories during demonstration flights.

replaced with boosters which are propelled with liquid oxygen-liquid hydrogen engines, similar to those carried on the Orbiter. The Liquid Fly-Back Boosters (LFBB) will be equipped with wings, and when jettisoned, would be able to fly automatically back to Kennedy Space Centre, landing rather like aircraft on a runway. The LFBBs will be safer, more reliable, less expensive and will perform than the old SRBs.

A NEW SPACE SHUTTLE

By 2005, NASA plans to have made a decision on the types of vehicles it would like to develop to replace the Shuttle. will

Meanwhile, NASA will continue to fly the Shuttle until the new generation spacecraft is available. If, for any reason, no new viable vehicle can be developed by 2015, the Shuttle system is perfectly capable of flying to 2030 according to some space engineers.

At this stage no one really knows what the new vehicle will look like. Many proposals have been put forward, perhaps the most famous of which is the VentureStar. Any vehicle that is going to replace the Shuttle will have to demonstrate safe, regular, routine space travel at a cost of less than ten times the cost to fly the Shuttle. A typical Shuttle mission budget can be more than $500 million. To produce such a vehicle will require the flight of several demonstrator vehicles to test the vital new technologies required to make space travel as routine as flying airliners.

The X-33 advanced technology demonstrator is one example. This delta-winged vehicle is a half-scale model of the planned 38m (124ft) long, 12.38 tonne, VentureStar. It will make 15 test flights, starting in around 2001 and will eventually fly to Mach 13 at an altitude of 91.5km (56.9 miles) by 2003. It will take off vertically under the power of new linear aerospike engines and land automatically

like an aircraft on an ordinary runway, after a journey of 1500km (932 miles) lasting 14 minutes.

The Aerospike engine represents the greatest advance in space technology, and also the biggest challenge. The engine operates like a rocket engine, consuming liquid oxygen and liquid hydrogen. However, its nozzle is made up of several curved nozzle ramps which form the structure of the engine. Multiple combustion chambers are mounted in rows at the forward ends of upper and lower ramps, which are made from copper sheet. Grooves are milled into the back side of the ramp, and a steel alloy sheet brazed on to

Right: The linear Aerospike engine will be crucial to the success of the X-33 and VentureStar programmes. Multiple combustion chambers are mounted in rows.

it to create passages through which liquid hydrogen is pumped to cool the ramp. The exhaust is propelled onto the ramps, and the emanating plumes adjust automatically with altitude, expanding as the ambient pressure decreases. This results in greater propulsive efficiency than a conventional rocket motor with a bell nozzle. The plume expansion automatically optimizes performance. Development and testing of this engine, however, has proved difficult and has delayed the programme.

In addition to the Aerospike engine, the advanced technologies of the X-33 include lightweight composite materials and new thermal protection systems. The composite materials were planned to be used to construct the tanks which store the liquid oxygen and liquid hydrogen propellants for the engines. However, production difficulties, including leaking tanks, may result in a move back to using heavier tanks – losing one of the benefits of the programme, and reducing the payload capability of the vehicle and its performance. Even before this setback, the speed had been reduced from a planned maximum of Mach 15.

ROTON

Another futuristic concept could be the piloted Roton booster, which will take off

Above: The latest configuration of the VentureStar has it carrying an external "piggy-back" payload bay.

Above: The Roton launcher will take off like a rocket and land like a helicopter. Here it is pictured on a test flight of its rotor system.

The company's name makes reference to the rotary pumping action of its RocketJet engine and free-spinning helicopter-style blades deployed for landing. The Roton lands gently and precisely using helicopter-style rotors which are simply folded flat against the vehicle's sides during powered flight. They are deployed during descent for piloted landing. The Roton RocketJet engine, with a thrust of 2225 kilonewtons and rotary altitude compensation, utilizes centrifugal force to spin kerosene fuel and liquid oxygen oxidizer out of dozens of small combusters, arranged in a ring at the rocket's base. The simple design eliminates the need for heavy and expensive turbopumps.

A Roton prototype was flown to test approach and landing within Earth's atmosphere using powered rotors. In making a series of "bunny-hop" tests, the craft eventually reached altitudes of between 1.52km (0.94 miles) and 2.44km (1.52 miles). A rocket-powered test flight of another test vehicle is planned for a later date, with a number of suborbital flights before Roton heads for orbit. The Roton will be piloted by two crew during all development and operational flights, housed in a cockpit towards the rear of the vehicle, rather than at the front. A flight-test team is being recruited from experienced graduates of military and civilian flight test programmes, who will be familiar in handling the latest high performance air and rotorcraft.

The company, which was experiencing a financial crisis in 2000, believes that the Roton will be safer to operate because it will be piloted. Airliners have multiple engines, so does the Roton. Airliners are able to return to land in the event of an emergency, an ability which the Roton will also possess. Although it is highly automated like an aircraft, the crew can quickly intervene in the event of a systems failure. Roton also carries one-fifth the amount of kerosene that a Boeing 747

like a rocket and land like a helicopter. The Rotary Rocket Company in California hopes to herald the arrival of routine commercial space transportation with its fleet of rapid-turnaround launchers, called Roton, which may become the world's first single-stage-to-orbit (SSTO), re-usable commercial boosters. Rotary has the start-up funds, but will need many more millions of dollars to convert the project into reality.

requires, and can land within a target of 30m (98ft) square, rather than the Boeing 747's long runway.

The Roton will initially be targeted at the satellite launcher market, but may later be used to carry passengers on sub-orbital or orbital hops into space. Other technology demonstrators such as the Hyper-X and X-34 are also being flown, but clearly the road to regular space travel will be much longer than planned.

THE FIRST COMET LANDER

The European Space Agency (ESA) plans to make the first landing of a spacecraft on a comet, soft-landing a 100kg (220lb) craft on Comet Wirtanen in 2012. The launch of the mission, called Rosetta, is scheduled for January 2003. Rosetta will also become the first craft to orbit a comet, depositing its small lander craft onto the surface of Wirtanen as it approaches the Sun and begins to shed material, forming the trademark tail of a comet.

To achieve the speed necessary to reach Wirtanen, Rosetta will require several gravity-assisted fly-bys of Earth and Mars. It will fly by Mars in May 2005 and then, in "Back-to-the-Future" mode, will fly by Earth in October 2005 and October 2007, before

Below: An artist's impression of the Rosetta spacecraft flying close to the nucleus of the comet Witanen in 2012, after a nine-year voyage through interplanetary space.

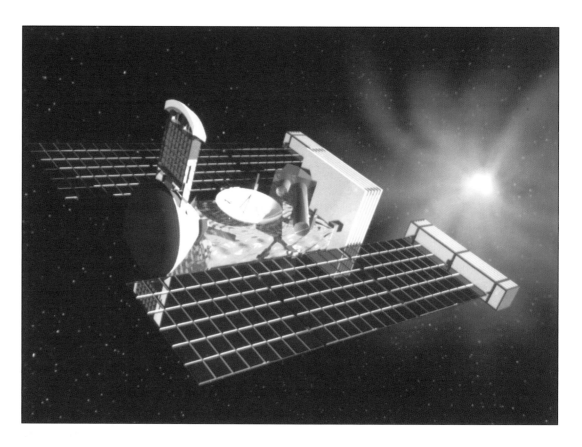

Right: A capsule deployed from the Stardust spacecraft will return to Earth in 2006 carrying the first samples from a comet for analysis by eager scientists.

it heads out towards a rendezvous with Wirtanen. En route, Rosetta will also include two asteroid visits, taking pictures of 4979 Otawara and 140 Siwa from a distance of 1000km (621 miles) in July 2006 and 2008 respectively. Siwa will be the largest asteroid yet explored by any spacecraft.

After the journey of 5.3 billion kilometres (3.29 million miles) around the solar system, Rosetta will make its first rendezvous with Wirtanen in November 2011, 675 million kilometres (419 million miles) from the Sun, in which sunlight is 20 times weaker than it is on Earth, and requiring the craft's huge solar panels – with spans as wide a football pitch – to work at maximum capacity to provide solar-generated electricity. By May 2012, Rosetta will eventually have closed to within 2km (1.24 miles) of Wirtanen's frozen nucleus and will enter orbit around it, sending back the most detailed images yet obtained of a comet body. A suitable landing site will be selected about one month after global mapping commences. The orbiter's small

lander will be released, and will make a soft touchdown at a speed of less than 1m (3.27ft) per second, to allow for the negligible gravitational pull of the tiny nucleus. To ensure that the lander does not bounce and disappear into space, an anchoring harpoon will be fired into the surface immediately on touchdown. The surface may even be porous and crusty like a meringue.

For the first time, scientists will be monitoring, at close quarters, the dramatic changes which take place as the comet hurtles towards the Sun at a speed of 46,000km/h (28,583mph). As a preview to Rosetta, another spacecraft, called Stardust, will return dust to Earth, collected from another comet, Wild 2, marking the first time that samples from another body in space, other than the Moon, will have been brought back to our planet. Another goal of the mission is also to gather small grains of interstellar matter.

Stardust, which resembles a giant catamaran, will make a 6.1km (3 miles) per second rendezvous with Wild 2 in January 2004, as the comet approaches the Sun on its long, lonely elliptical solar orbit. The 100km (62 mile) fly-by of the comet will be part of a 10-hour flight through Wild 2's coma of 200,000km (124,274 miles) width. The comet consists principally of material being shed from its nucleus – which is basically a dirty snowball with a rocky core – which forms a tail as the comet is heated while passing close to the Sun.

The key to the success of the Stardust mission is ultra-low density silica aerogel, a sticky material which will be mounted like a communications dish on a mast extended from a sample return capsule (SRC) mounted on the spacecraft. As Stardust passes Wild 2, coma dust in the 1–100 micron range will be captured on one side of the Aerogel dust collector. Stardust will have already collected interstellar dust striking the spacecraft at a velocity of 30km (18.64 miles) per second. This material will be collected on the opposite side of the dust collector. It is planned to collect over 100 particles in the 0.1–1 micron size. The objective of the Wild 2 encounter is to recover more than 1000 particles larger than 15 microns in diameter, as well as volatile molecules.

VItal to the mission's success, of course, is the return of the samples to Earth in January 2006, the other innovative achievement for the 303kg (667lb) spacecraft. The relatively tiny 25kg (55lb) SRC will be released from the main spacecraft around three hours before re-entry into Earth's atmosphere, while the main craft performs a diverting manoeuvre to force it to fly past Earth. The SRC will encounter Earth's atmosphere at a velocity of 12.8km (7.95 miles) per second, at an altitude of 125km (77.7 miles). Protected by an aeroshell heat shield which will dissipate 99 per cent of the kinetic energy, the capsule will deploy its parachute 10 minutes later, landing within a "footprint" measuring around 60km (37 miles) by 6.5km

Left: This should have been the scene in late 1999: the Mars Climate Orbiter and the Mars Polar Lander exploring the "Red Planet". However, both vehicles failed.

(4 miles) in Utah. After recovery, the capsule and its precious samples will be transferred to the Planetary Materials Curatorial Facility at the Johnson Space Centre in Houston, Texas, for onward distribution to investigators worldwide.

MAN ON MARS?

Apart from witnessing the first men walking on the Moon, nothing has captured the imagination of the public more than the landing of the Mars Pathfinder and its Sojourner rover in July 1997. Mars has always fascinated mankind, and one day humans will no doubt walk upon its ruddy-coloured surface. Quite besides the huge technological challenge of sending humans to Mars, the major hurdle is the enormous cost. It is now simply too costly to fly into space, and a mission to Mars, for the sake of it, will not be acceptable to taxpayers until it is significantly cheaper. In addition, the technical hurdles of supporting a manned mission are enormous. If there is ever going to be a manned Mars mission, it is likely to be a relatively modest affair.

In 1994, NASA had developed a launch strategy for a possible expedition to Mars, to be launched in 2011, and involving three launches of a huge new booster. But even this plan was considered too ambitious, both technically and financially. NASA scaled down the proposed baseline mission by using smaller boosters. However, twice the number of launches would be required to assemble the components in Earth orbit for despatch to Mars.

The proposed scheme was as follows. An inflatable TransHab module would be launched to Mars with the Cargo 1 craft, to make landfall in the prime landing zone. This module would serve as a living and experiment quarters for the Mars crew when they arrived – assuming a safe and accurate landing. The crew would land in a separate spacecraft, then connect it to TransHab. The upper part of the manned Cargo vehicle would later take off, following the proposed 500-day exploration mission of the planet's surface, and then dock with the return craft which would have been placed in Mars orbit earlier.

There is another important issue in the human exploration of Mars. The Apollo astronauts were three days away from Earth. However, a Mars crew would face similar

Right: A NASA depiction of a possible Mars base, comprising habitation modules which would be followed later by a manned lander.

Left: An impression of a very
futuristic mission, exploring
the Martian moon, Phobos.

hazards with no immediate return to Earth in prospect. Rather like the first ocean voyagers, loss of ships have to be expected but are unlikely to be tolerated by the public.

INTELLIGENT SPACECRAFT

Flying far away from Earth, beyond reasonable range of communications, will increasingly require "intelligent spacecraft" which can resolve on-board problems normally overseen by mission control.

Deep Space 1 (DS 1), which was launched in July 1998, is the first scheduled mission in NASA's New Millennium programme, which is designed to test and validate cutting-edge technology for systems and instruments on board future NASA science spacecraft. It has already accelerated spacecraft automation technology by at least ten years, demonstrating the most advanced spacecraft advanced intelligence software yet developed. It is the nearest thing yet to HAL 9000, the main computer from the landmark science-fiction story, 2001: A Space Odyssey, written in 1968 by Arthur C.Clarke and later made a film. The robotic DS1 spacecraft carries no crew and, at 945kg (2083lb), is much smaller than Clarke's spacecraft, but its computer artificial intelligence programme, known as Remote Agent, shares the same basic goals of operating and controlling a spacecraft with the minimum of human intervention.

Three functions of the Remote Agent – high-level planning and scheduling, model-based fault protection, and smart executive – work together to enable it to operate a spacecraft autonomously. The high-level planning and scheduling function, or The Planner, of the Remote Agent constantly looks ahead in the schedule for several weeks of mission activities. It is chiefly concerned with scheduling spacecraft activities and distributing resources such as electrical power. If one part of a spacecraft was to perform differently than expected during the mission, the craft would be able to detect this, and change software models and algorithms to self-adapt. The fault-protection portion of the Remote Agent, called "Livingstone", functions as the mission's virtual chief engineer. If something goes wrong with the craft, Livingstone – named after Sir David Livingstone, who was concerned both with exploration and the health of explorers – uses the computer model of how the spacecraft should be behaving to diagnose failure and suggest recoveries.

The third part of the Remote Agent software, the Smart Executive, acts like the executive officer of the mission, issuing

SHEN ZOU

The Chinese Shen Zou spacecraft may make its first manned flight in 2001. It resembles a Russian Soyuz craft, with a service module, below, a central crew capsule, and an orbital module with a docking collar.

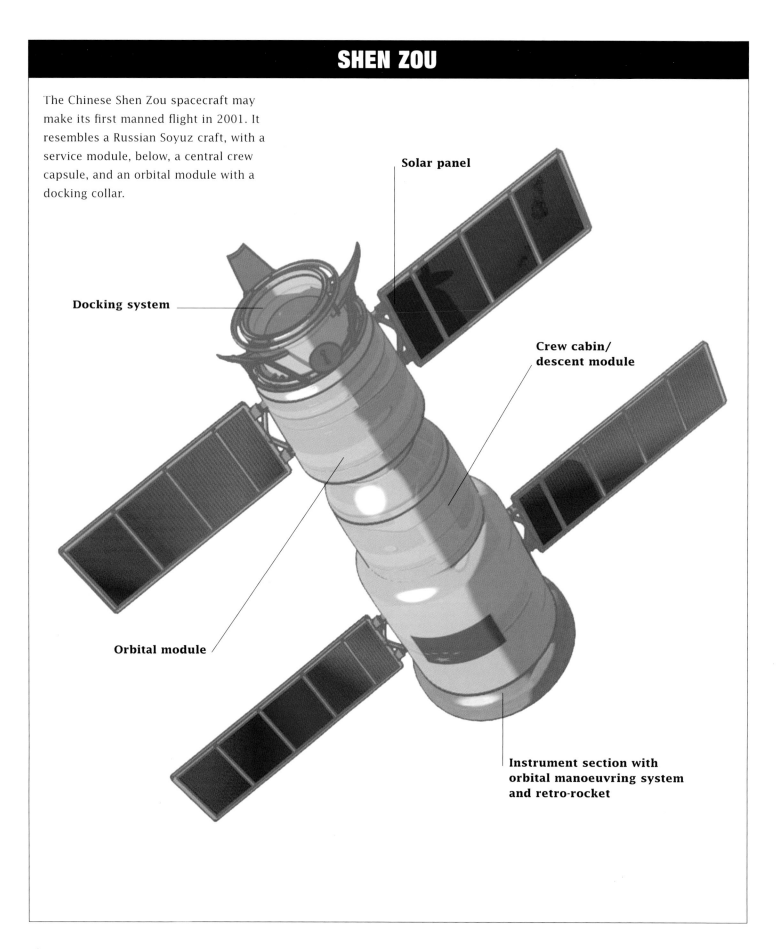

Solar panel

Docking system

Crew cabin/ descent module

Orbital module

Instrument section with orbital manoeuvring system and retro-rocket

WHERE ARE WE GOING?

Regarded in terms of the history of aviation, the field of space exploration since the launch of Sputnik 1 has only reached the year 1944, 41 years after the first powered flight at Kitty Hawk. Today, airliners can fly around the world and carry about 500 people. Some air passengers fly at twice the speed of sound. What will be happening in space in another 41 years time? Space hotels, a human base on Mars, star travel? As a warning to those who have hopes of tourism in space, we should look back to some of the things that were meant to have happened by now, judging from what was predicted only 30 years ago when Man first set foot upon the Moon. Two perfectly rational predictions at the time were that men would land on Mars in 1985 and that the President of the US would travel to a space station in 2000. That these things did not happen is due mainly to cost. If the US spent 20 times the amount it spent on Apollo, it may well have got to Mars. As it turned out, the US eventually spent five times Apollo's money on the Space Shuttle. If, at the same time, a space station together with the Space Shuttle were developed as an integral part of the Mars-landing goal, then we might have been closer to achieving it. A space station was eventually given the go-ahead 10 years later than was hoped, and is going to be 10 years late in completing its construction in orbit. Space is just too expensive, and the first task in the 21st century will be to make it significantly cheaper.

If great advances are going to be made, access to space will have to be as routine as air travel – and as safe. These things are not yet possible, and are unlikely to be

Above: NASA once made plans for a spaceplane that could fly from the US to Australia in two hours. However, the National Aerospace Plane project has practically disappeared from consideration.

achieved in the next 20 years, judging by NASA's admission that the Space Shuttle could still be flying in 2015 or later. Budgetary considerations, regular access and safety will be controlling factors in the future of human space travel. Moreover, quantum leaps in technology will have to be made before we move into a new space era. Computers and communications have improved in an extraordinary – though evolutionary – manner, but the rocket engine of today is fundamentally still the same as it was in 1957, when the first satellite, Sputnik 1, was launched.

general commands to fly. The Executive has to be able to execute the plans that are produced by the Planner and Livingstone. If the Planner had to concern itself with every single detail, it would be hard pressed to produce any plan at all. So, the Executive takes care of the details.

THE HYDROBOT

One of the applications of the Remote Agent could be on the mission to launch a mini-submarine into the waters of Jupiter's moon, Europa. NASA plans to launch a $250 million Cryobot spacecraft in around 2010 to penetrate the ice of Europa, and deploy a miniature submarine, called a Hydrobot, into what are believed to be the waters beneath. This audacious mission is part of scientists' insatiable quest to find signs of life elsewhere in the solar system. Images from NASA's Jupiter-orbiter, Galileo, have shown Europa to be covered with smooth white and brown-tinted ice. The ice has been cracked and moved by the enormous tidal effects of Jupiter's gravity like a jig-saw coming slightly apart. Scientists believe this has

Above: A German concept of a two-piece spaceplane, comprising a booster stage which would return to Earth like an aircraft, and a "piggy-back" spaceplane.

been caused by warm ice and the action of water beneath the surface. Scientists have calculated that, based on radioactive heat sources and heat transported by conduction, Europa could have a water mantle 100km (60 miles) thick, of which 50km (30 miles) could be liquid water. If there is water present, so the theory goes, parts of Europa's oceans may be warm enough to support primitive life forms.

The Cryobot/Hydrobot mission is the brainchild of NASA's Jet Propulsion Laboratory in Pasadena, California. The 1.22m (4ft) long Cryobot penetrator-melter probe, similar to the Philberth Probe used by polar explorers, will have a heated tip which will allow it to melt a hole in the ice at a rate of about 0.9m (2.95ft) a day. The heat would be provided by 16 radioisotope thermo-electric generators providing 4kW of thermal energy. Europa's ice is thought to be at least half a mile thick. Once the Cryobot has reached the water – which could take weeks or even months – it will open a hatch and deploy an even smaller, self-propelled Hydrobot probe which will "pop out" and snake through the water, using a powerful search light, and cover a

distance of about a 0.8km (0.5 miles). A miniature camera and chemical sensors will search for life. These will be contained in a minute laboratory which will determine whether the ocean exhibits a set of required sources for life – liquid water, carbon, nitrogen, phosphorous and sulphur. The craft may also perform DNA fluorescence checks. The Hydrobot will also be equipped with a hydrophone to listen for volcanism, a thermometer and sonar imager, and may even return samples to the Cryobot for further on-board analysis.

The technology and flight of the Cryobot-Hydrobot mission will be developed and tested extensively in a sub-glacial lake in Antarctica, which scientists believe resembles conditions on Europa. Lake Vostok in the Russian territory of eastern Antarctica contains water chemistry and lacustrine sediments, possibly heated volcanically, making it a suitable habitat for simple life-forms, and mimicking conditions that some scientists believe exist on Europa. Deep vents in the oceans of other parts of the Earth have been found to contain bacteria, tube worms and other organisms.

The Lake Vostok project, in which the Cryobot/Hydrobot mission will be "flown" to demonstrate and test the technology required for the actual mission, will complement a planned systematic, step-by-step assault on Europa, culminating in the Cryobot/Hydrobot mission. The first mission has been dubbed the "Ice Clipper". This will be launched to orbit Europa at a height of 100km (60 miles) and conduct a radar-sounding of the ice moon to determine the existence of water. The Ice Clipper may also deploy impact penetrators to determine the relief and chemical composition of the surface of Europa.

The next mission will be launched in 2005 to land a Cryobot on the surface of Europa to conduct the first shallow melt of 457m (1500ft) into the ice as a technology demonstrator. The Cryobot will also take images and measurements of the surface and test for the existence of trace organics that may imply a metabolic pathway, such as ethanol, butanol and acetate. If this test is successful, then the Cryobot/Hydrobot mission would be launched in about 2010. Flying to a distant moon, landing safely on the surface, and penetrating the ice is a scheme that sounds on the verge of being far-fetched – but that was what people once said about man on the Moon.

Below: A proposed mission to Europa, Jupiter's oceanic moon covered by an ice sheet, would send a Cryobot craft to penetrate the ice sheet before deploying the Hydrobot explore into the waters.

GLOSSARY

AAP Apollo Applications Programme.

Ablation The erosion of a solid body by a high-temperature gas stream moving with high velocity, e.g. a re-entry vehicle's heat shield which melts or chars under the effects of air friction.

Abort To cancel or cut short a mission.

AFB Air Force Base (US).

AFRSI Advanced Flexible Reusable Surface Insulation (Space Shuttle orbiter).

ALT Approach and Landing Test.

Aphelion The point in a solar orbit which is farthest from the Sun.

Apogee The point in a terrestrial orbit which is farthest from the Earth.

Arianespace A European commercial consortium.

ASCS Automatic Stabilization and Control System (Mercury capsules).

Asteroid A small planetary body. Many thousands of them orbit the Sun between the orbits of Mars and Jupiter.

ASTP Apollo-Soyuz Test Project.

Astronaut A person who flies in space, whether as a crew member or passenger.

Atmosphere The envelope of gases surrounding the Earth or any other planet.

Attitude Orientation of a space vehicle as determined by the relationship between its axes and some reference plane, e.g. the horizon.

Ballistics The science that deals with the motion, behaviour, appearance or modification of missiles acted upon by external forces.

Booster The first stage of a missile or rocket.

CDR Commander.

Chaff Metallic foil ejected by a re-entry module to enhance its radar image.

CM Command Module. The compartment of a spacecraft which contains the crew and main controls.

Coma A spherical cloud of gas and dust surrounding the nucleus of a comet.

Combustion chamber The chamber in a rocket where the fuel and oxidant are ignited and burned.

Comet A body of small mass but large volume, compared to a planet, orbiting the Sun, often developing a long luminous tail when close to the Sun.

Control rocket A vernier or other rocket used to control the attitude, or slightly change the speed of a spacecraft.

Cosmonaut The Russian term for an astronaut.

COSTAR Corrective Optics Space Telescope Axial Replacement unit (HST).

Crossrange Manoeuvres to the left or right which can be made by the Space Shuttle orbiter in order to align itself for an emergency landing.

Cryogenic A rocket fuel or oxidant which is liquid only at very low temperatures, eg. liquid hydrogen.

CSM The Command and Service Modules together.

Doppler effect The apparent change in frequency of vibration (i.e. pitch) of sound, light or radio waves, due to relative motion between the source and the observer.

Drag The resistance offered by a gas or liquid to a body moving through it.

Drogue A small parachute used to slow and stabilize a spacecraft returning to the atmosphere, usually preceding deployment of a main landing parachute.

EDO Extended Duration Orbiter, Shuttle orbiter with more fuel cells.

Elints Electronic intelligence satellites.

EOR Earth Orbit Rendezvous.

EOS Earth Observation System (series of satellites, NASA).

Equatorial orbit An orbit in the plane of the equator.

ESA European Space Agency.

Escape velocity The precise velocity necessary to escape from a given point in a gravitational field.

ET External Tank (Space Shuttle).

EVA Extra-Vehicular Activity.

Exhaust velocity The velocity of the exhaust leaving the nozzle of a rocket motor.

FGS Fine Guidance Sensors (HST).

Fly-by Space flight past a heavenly body without orbiting.

FOC Faint Object Camera (HST).

FOS Faint-Object Spectrograph (HST).

FRCI Fibrous Refractory Composite Insulation (Space Shuttle orbiter).

FRSI Felt Reusable Surface Insulation (Space Shuttle orbiter).

Fuel cell A cell in which chemical reaction is used directly to produce electricity.

g The symbol for the acceleration of a freely moving body due to gravity at the surface of the Earth.

GDL Leningrad Gas Dynamics Laboratory, est. 1928, USSR.

GEO Geostationary orbit. A circular orbit in which a satellite moves from west to east at such a velocity as to remain fixed above a particular point on the equator.

Geodesy The science of the Earth's shape.

Geophysics The physics of the Earth.

GHRS Goddard High-Resolution Spectograph (HST).

Gimbal A mechanical frame for a gyroscope or power unit, usually with two perpendicular axes of rotation.

GIRD Group for the Study of Reaction Propulsion, est. 1931, USSR.

Glass cockpit New Shuttle orbiter cockpit in which old gauges and instruments are replaced with 11 full-colour flat panel screens.

GLONASS Global Navigation Satellite System (Russia).

GPS Global Positioning System.

Gravitational field The region of space surrounding a body in which another body experiences a force of attraction.

Gravity The force that attracts a body to the centre of the Earth or other physical body having mass.

Gyroscope A device consisting of a wheel so mounted that its spinning axis is free to rotate about either of two other axes perpendicular to itself and to each other. Once set in rotation the gyro axle will maintain a constant direction regardless of the fact Earth is turning under it.

Heat shield A device which protects people or equipment from heat such as a shield in front of a re-entry capsule.

HRSI High-temperature Reusable Surface Insulation (Space Shuttle orbiter).

HSP High-Speed Photometer (HST).

HST Hubble Space Telescope.

Hypergolic A term applied to an oxidant and a fuel which ignite spontaneously with each other.

ICBM Intercontinental Ballistic Missile.

IGY International Geophysical Year.

ILS International Launch Services, a commercial satellite-launching organization.

IMAGE Imager for Magnetopause-to-Aurora Global Exploration satellite (NASA).

Inclination The angle at which an orbit crosses the equator.

Ion An atom that has lost or acquired one or more electrons.

Ion engine A rocket engine, the thrust of which is obtained by the electrostatic acceleration of ionized particles.

Ionosphere The region of the Earth's upper atmosphere which reflects or absorbs radiowaves.

IRBM Intermediate Range Ballistic Missile.

ISS International Space Station.

KSC Kennedy Space Centre, Cape Canaveral, Florida.

Launch window An interval of time during which a space vehicle can be launched on a given mission.

LEO Low-Earth Orbit.

LES Launch Escape System.

LFBB Liquid Fly-Back Boosters (improved Space Shuttle).

LIDAR Light-Detection and Ranging instruments on the Picasso Earth satellite (NASA).

LM Lunar Module.

LOR Lunar Orbit Rendezvous.

LOX-LH Liquid oxygen-liquid hydrogen.

LRSI Low-temperature Reusable Surface Insulation (Space Shuttle orbiter).

LRV Lunar Roving Vehicle.

Mach The ration of the speed of a vehicle (or of a liquid or gas) to the local speed of sound.

Magnetic field A region of variable forces around magnets, and magnetic materials.

Magnetometer An instrument measuring magnetic forces (particularly Earth's).

Magnetosphere The region of space surrounding Earth which is dominated by the magnetic field.

Max Q Maximum dynamic pressure: the point during launch when the vehicle is subjected to its greatest aerodynamic stress.

MDU Multifunction Display Units (Shuttle orbiter glass cockpit).

Micrometeoroid Meoteroid less than 250th of an inch in diameter.

MSG Meteosat Second Generation (ESA).

NASA National Aeronautics and Space Administration, est. 1958, USA.

GLOSSARY

NEAR Near Earth Asteroid Rendezvous asteroid orbiter (NASA).

NICMOS Near-Infrared Camera and Multi-Object Spectrometer (HST).

Nose shroud A cover on the nose of a rocket or spacecraft which jettisons before insertion into orbit.

OMS Orbital Manoeuvring System (Space Shuttle orbiter).

Orbit The curved, usually closed course of a planet, satellite etc.

Orbital period The time taken by an orbiting body to complete one orbit.

Orbital velocity The velocity necessary to overcome the gravitational pull of the Earth and so keep a satellite in orbit.

OSO Orbiting Solar Observatory.

OST Orbital Solar Telescope on Salyut 4 space station.

OTA Optical Telescope Assembly (HST).

Oxidants, oxidisers Chemicals used for combining with fuels in rocket engines to enable combustion to be independent of the atmosphere.

Parking orbit Orbit in which a space vehicle awaits the next phase of a planned mission.

Payload Useful cargo.

Perigee The point in a terrestrial orbit which is nearest to the Earth.

Perihelion The point in a solar orbit which is nearest to the Sun.

Photovoltaic cells Crystalline wafers called solar cells which convert sunlight directly into electricity without moving parts.

Pitch Movement of a spacecraft about an axis (Y) which is perpendicular to its longitudinal axis: degree of elevation or depression.

PLT Pilot.

Polar orbit An orbit which passes over the poles.

Radio telescope A radio receiving station for detecting radio waves emitted by celestial bodies or space vehicles.

RCS Reaction Control System (Space Shuttle orbiter).

Re-entry The re-entry of a space vehicle into the atmosphere.

Retro-rocket Rocket fired to reduce the speed of a spacecraft.

RMS Remote Manipulator System, sophisticated robotic arm on the Space Shuttle orbiter.

RNII Scientific Rocket Research Institute (successor to GIRD).

Roll The rotational movement of a vehicles about a longitudinal (X) axis.

Samos Satellite and Missile Observation System.

Satellite A natural or artificial body moving around a celestial body.

SCORE Signal Communications Orbit Relay Experiment.

SM Service Module. The part of a spacecraft which usually carries a manoeuvre engine, electrical supply, oxygen and other consumables external to the descent module. Discarded prior to re-entry.

Solar cell A cell that converts sunlight into electrical energy. The light falling on certain substances (e.g. a silicon cell) causes an electrical current to flow.

Solrad Solar Radiation Satellite.

Sounding rocket A research rocket used to obtain data from the upper atmosphere.

SRB Solid Rocket Booster.

SRC Sample Return Capsule.

SSME Space Shuttle Main Engine.

SSTO Single-Stage-to-Orbit, a new type of booster.

STIS Space Telescope Imaging Spectograph (HST).

Sub-orbital Not attaining orbit, i.e. a ballistic space shot.

Sun-synchronous orbit An orbit in which a satellite passes over the same place on the equator at the same time each day.

Telemetry The system for radioing information including instrument readings and recordings from an air or space vehicle to the ground.

Thrust Propulsive force. Measured in lb, kg, or Newtons.

TPS Thermal Protection System (Space Shuttle orbiter).

Trajectory The flight path of a projectile, missile, rocket or satellite.

Trans-lunar injection The re-ignition of Apollo's engines after a period in parking orbit, in order to set it on course to the Moon.

Transponder Radio equipment that receives a signal, modulates it, and re-transmits it at a different frequency.

VAB Vehicle Assembly Building (Space Shuttle) at KSC.

Vernier Rocket engine of small thrust used for fine adjustments in velocity and trajectory.

VfR Verein für Raumschiffahrt (Society for Space Travel) est. 1927, Germany.

Volcam Volcanic Ash Mission (NASA).

WFPC Wide Field/Planetary Camera (HST).

Yaw The rotation of a vehicle about its vertical (Z) axis, i.e. to a different azimuth.

INDEX

ACKNOWLEDGEMENTS

The author gratefully acknowledges references to statistical information provided from the publications listed below, and editions of the *TRW Space Log*:

The Viking Rocket Story by Milton Rosen, Panther, 1957
Missiles and Rockets, by Kenneth Gatland, Blandford, 1975
The Soviet Manned Space Programme by Phillip Clark, Salamander, 1988
The Soviet Reach for the Moon, by Nicholas Johnson, Cosmos Books, 1994

Spacecraft and Boosters, by Kenneth Gatland, Iliffe, 1964
Spacecraft and Boosters, by Kenneth Gatland, Illife, 1965
Soviet Rocketry, by Michael Stoiko, David and Charles, 1970
Solar System Log, by Andrew Wilson, Jane's, 1987
Guinness Space Flight The Records, by Tim Furniss, Guinness Books, 1985
Manned Spaceflight Log, by Tim Furniss, Jane's, 1983 and 1986
Space Shuttle Log, by Tim Furniss, Jane's, 1986
Rockets, Missiles, and Men in Space, by Willy Ley, Signet Books, 1968

PICTURE CREDITS

Associated Press: 44 (t), 206.
Chrysalis Pictures: 11, 16, 17, 18, 52, 115 (both), 116, 163.
Genesis Space Photo Library: 8, 9, 21 (both), 22 (l), 23, 25 (r), 26-27, 31, 33, 34 (t), 36, 37, 38, 39 (both), 42, 43 (both), 44 (b), 45, 46, 47 (t), 47 (b) (NASA), 48-49, 54, 56 (both), 57 (both), 58, 59, 63, 65 (NASA), 66-67, 68, 70 (both), 73, 75, 76, 77, 78 (both), 82, 84, 87, 88-89, 90, 91, 92, 93, 95, 97 (both), 98 (l), 104, 107,

108, 110 (r), 111 (l), 113, 118-119, 120, 121, 122, 124, 125, 127, 130, 131, 134, 135 (both), 136, 138, 139 (both), 142, 143, 145 (both), 146-147, 148, 149, 156, 164, 165, 167 (both), 168-169, 170, 172, 173, 174-175, 177, 180, 182, 183, 185 (t), 186, 187, 190, 192, 193, 195 (both), 196, 197, 198, 199, 200, 202, 203, 204, 208, 215, 218, 220, 221, 223, 225, 226, 227, 228-229, 230, 231, 232, 233, 234, 235, 236, 237, 238, 239, 240, 241, 242,

243, 245, 246, 247. Novosti: 15, 30, 51, 53 (t), 209, 210, 212.
The Society for Co-operation in Russian and Soviet Studies: 53 (b).
Frank Spooner Pictures: 19.
TRH Pictures: 6-7 (Bundesarchiv), 25 (l) (NASA), 29, 111 (r) (Lockheed).

Artworks:
De Agostini UK: 205, 207, 213, 214, 222, 224.

Chrysalis Pictures: 22 (r), 32, 34 (b), 40-41, 50, 55, 60, 62, 64, 69, 71, 74-75, 79, 81, 83, 86, 94, 96, 98 (r), 99, 100, 101, 102, 105 (both), 106 (both), 109, 110 (l), 112, 114, 126, 128, 129, 132, 133, 137, 150-151, 152, 153, 154-155, 157, 158, 160-161, 162-163, 178-179, 181, 184, 185 (b), 189, 191, 194, 211.
Martin Woodward (tecmedi): 12, 13, 14, 24, 80, 123, 141, 144, 171, 188, 244.